BIOLOGICAL AND ENVIRONMENTAL HAZARDS, RISKS, AND DISASTERS: KEY AREAS FOR STUDY

BIOLOGICAL AND ENVIRONMENTAL HAZARDS, RISKS, AND DISASTERS:

KEY AREAS FOR STUDY

Edited by

Dr. Esha Rami

Delve PUBLISHING

www.delvepublishing.com

Biological and Environmental Hazards, Risks, and Disasters: Key Areas for Study

Dr. Esha Rami

Delve Publishing
224 Shoreacres Road
Burlington, ON L7L 2H2
Canada
www.delvepublishing.com
Email: orders@arclereducation.com

ISBN: 978-1-77469-872-3 (Hardcover)

ABOUT THE EDITOR

Dr. Esha Rami is presently working as an Assistant Professor, in the Department of Life science, Parul Institute of Applied Science, Parul University, India. She Did her Post graduated from Ganpat University, Ph.D. in biotechnology from Hemchandracharya North Gujarat University, in the year 2015. She has authored a number of national and international publications in reputed journals.

TABLE OF CONTENTS

LIST OF FIGURES

ABSTRACT

This book provides an integrated outlook on key impacts on the Earth's biosphere. In particular, natural factors such as disease, deforestation, algal blooms, insect and animal extinction, volcanic eruptions, asteroids and land degradation have considerable impacts on humans, and most risks, hazards and disasters are caused by them. From microbes to celestial bodies, this volume provides an in-depth analysis of potential threats. The opinions on the possible effects of these factors on humans and the environment have been compiled. Readers will gain valued insights into damages, causality, economic impacts, preparedness, and mitigation measures from contributions by experienced biological, environmental, health, ecological, wildlife, and physical health scientists.

Natural events are considered hazards when they can possibly harm individuals or cause property destruction, and socioeconomic disruption. The areas of natural hazards are based on natural processes, such as the shifting of tectonic plates, weather system impacts, and the presence of streams and slopes (for example, that could produce avalanches). However, processes like environmental degradation, urbanization, and climate change may also affect where natural hazards occur, how often they occur, and their intensity levels.

PREFACE

This book on "Biological and Environmental Hazards, Risks, and Disasters" offers a comprehensive outlook on key effects on the Earth's biosphere. These factors are known as risk drivers; algal blooms, diseases, insect/animal extinction, and deforestation, many of which are caused by humankind. From microbes to celestial objects, thebook provides an analysis of these potential threats. Viewpoints from both natural science and sociologies are shared by biologists, ecologists, and nature researchers among other experts.

Tropical cyclones—also called hurricanes in the Atlantic Ocean, typhoons in the Northern Pacific Ocean, and cyclones in the Indian Ocean-can bring extreme winds, storm surges, and intense rainfall, and cause secondary hazards like landslides. Each hazard frequently causes a subset of others. A progression of triggering relations can cause a domino impact, for example, the tsunami caused by seismic tremor following the 2011 Japan nuclear crisis.

Hazards happen at various intensities (or magnitudes) across various time scales also called temporal scales. Scientists discuss the occurrence of hazards of various intensities with regard to probabilities and return periods, which are also referred to as recurrence intervals. In general, the intensity of a hazard is proportional to the length of the return period, which indicates how rare the hazard is. Due to these long return durations, some regions may not have a memory of the expected dangers of high-intensity hazards. This was witnessed during the Mt Pinatubo 1991 eruption- the second biggest volcanic emission of the twentieth century- which dislodged 20,000 native people residing in its foothills and set off enormous landslides called 'lahars', which impacted individuals for a considerable length of time after the emission. Hazards can likewise happen at various geological (spatial) scales. For example, the occurrence and effect of storms will in general be localized, while droughts can happen over thousands of kilometers. The distinctions between extensive and intensive risk are discussed in this book.

Different nations face multiple threats. As a result, it is critical to take into account the risk associated with the whole range of hazards that could harm people or assets. Besides, there have been a few instances whereby the whole spectrum of hazards has not been regarded; for example, following the destructive 2004 Indian Ocean tsunami, a couple of houses in Aceh, Indonesia, were rebuilt in flood-prone zones, exposing families to potential hazards in the future.

Additionally, hazards can interact; Typhoon Yunya accompanied the explosion of Mt. Pinatubo in the Philippines in 1991, soaking the ash fall from the volcano in rain. The tremendous weight of the wet ash made rooftops to collapse, causing approximately 300 deaths in the aftermath. This illustration demonstrates that the interaction of hazards can have a greater overall impact than when they had occurred independently, which has significant effects on risk assessment. The identification of relevant hazard(s) and the gathering of hazard-related data are key stages in hazard assessment. The next step typically involves acquiring a variety of data related to the hazards upon identification. The most crucial information involves the date, extent, and level of intensity of historical events.

Chapter

1 A BRIEF INTRODUCTION TO ENVIRONMENTAL HAZARDS, RISKS, AND DISASTERS

CONTENTS

1.1 INTRODUCTION

Traditionally, meteorology, hydrology, geology, geography, and engineering, as well as human geography, sociology, psychology, and health and safety, have been included in the study of environmental hazards. To investigate environmental hazards and disaster management, a multidisciplinary and integrated strategy has been employed over the past two decades. This was partially due to the altering geography of the globe. The end of the Cold War between the West and the East and the collapse of the Berlin Wall in 1989 led to political, economic, and social stability in the West and the expansion of economic powers in the East. Rapid globalization has brought the West and the East closer together, and crises such as global warming, pandemic influenza, and international terrorist attacks now affect every person on Earth. Local issues require global solutions. The 2010's food insecurity in Russia and China, the 2011's political unrest in North Africa, and the 2011's earthquake-tsunami-radiation catastrophe in Japan had a significant effect on virtually everyone. Natural and man-made disasters are gaining in popularity as a subject matter. If the 2001 terrorist attack on the Twin Towers was still viewed as a singular anti-American event, it could be argued that the Boxing Day tsunami in Indonesia in 2004 was the event that ultimately awoke people (Jones, 2015).

The public's perceptions of environmental hazards have undergone a radical transition. The harrowing images of wounded victims and distraught families weeping because their loved ones were buried by falling debris during an earthquake captured the attention of the media. Due to the internet, people no longer presume that environmental hazards only affect those who reside far away or do not know them personally. Hurricane Katrina in 2005, the earthquakes in Pakistan and Kashmir in 2005, the July 7 bombings in London in 2005, the earthquake in Haiti in 2010, the snowstorms in Europe in 2010, the floods and cyclone Yasi in Australia in 2011, the earthquake in Christchurch in 2011, and the disaster in Japan (earthquake, tsunami, and nuclear plant meltdown in 2011) are examples for testing integrated disaster management strategies. Even if the hazards are unanticipated, it is possible to avoid catastrophes. Dangers, vulnerability, and risk are crucial aspects of disaster management. For this issue to be resolved, the following questions must be addressed: Is the environment growing increasingly hazardous? Are natural disasters growing worse? What is a catastrophe, exactly? Are threats becoming more hazardous for us? Is setting a risk threshold possible? Our susceptibility is increasing or decreasing. How is susceptibility evaluated? Can hazards or natural calamities be avoided? How effective is our disaster management? What function will future disaster management serve? Certain terms require definition. When nature is at its most extreme, natural occurrences occur, but they do not harm inconvenience, or murder residents. Natural or man-made events or occurrences with the potential to cause property damage, disruption of normal activities, loss of life, or environmental impact constitute hazards. A catastrophe is a very hazardous peril that has already occurred, affecting a large number of people and activities extremely and resulting in widespread losses of people, property, or the environment that exceed the capacity of the community to cope. A catastrophe is defined as a calamity that causes severe, immediate damage to people in addition to abrupt, unforeseen consequences (Paul, B. 2011).

Because of the environment, humans have access to resources and opportunities (such as water, oxygen, fire, minerals, and timber). When nature's disequilibrium exceeds the natural fluctuation line, however, extreme environmental occurrences, hazards, or disasters may occur. Volcanoes are a geographical hazard, whereas avian flu is a non-geographic hazard. Seasonal variations may present some dangers. Over the Western North Atlantic Ocean Basin, for instance, hurricanes may only occur during the summer. Some hazards can occur throughout the year (Jones, 2015).

Figure 1.1: The components of risk.

Source: Peduzzi P. The Disaster Risk, Global Change, and Sustainability Nexus. *Sustainability*. 2019; 11(4):957. https://doi.org/10.3390/su11040957 [accessed 12 Jun, 2023]

Landslides, for instance, can occur at any time of year. Some hazards (like earthquakes) are incidental, while others (like terrorist acts) are deliberate. Hazards' effects can be both tangible (such as the loss of lives) and intangible (such as stress and post-traumatic disorder, harm to society's integrity), direct (such as physical damage to buildings and contents), and indirect (such as losses of business, revenue, sales, and employment) (Smith, 2001). However, if there is more than one hazard, the incidence, impact, and management of risks would be complicated. The risk coupling may manifest in one of the following three ways: (a) two or more hazards occurring concurrently at different locations within the same nation that require the same resources for response and recovery (for example, the earthquake and flood in South China in May 2008 in Sichuan and Wenchuan); (b) two or more hazards occurring consecutively in the same location where the impacts and damages may be exacerbated and the recovery and reconstruction processes may be slowed (for example, the Japan disaster in 2011, magnitude 9.0), and (c) two or more hazards occurring simultaneously at different locations within the Each threat has an effect that cannot be quantified (Paul, 2011).

On the one hand, the aggregate damage can be less severe if two threats happen at different times. On the other hand, several threats would significantly tax the personnel and equipment available for emergency response, worsening the issue. How can the effects of dangers be estimated? Can we rapidly determine the number of people injured or slain, as well as the number of homes that have been completely or partially destroyed? The upheaval and disruption of social order that follows a calamity make it more difficult to account for losses and fatalities. For instance, after an earthquake, some people may still be alive but buried beneath the debris, while others may have escaped, died on the scene but were not reported to authorities, or were unfortunate tourists. It is crucial to estimate the approximate number of casualties (none, slight, moderate, serious, very serious, collapse), structural damages (no damage, partially damaged, destroyed), and financial losses (direct cost of damage, replacement cost of buildings and infrastructure, economic cost of restoring basic services) to request and deploy the necessary resources and emergency services. However, the financial consequences of a catastrophe extend beyond the short- and long-term losses of businesses to include the long-term impact on the public's confidence in and reputation for secure investment areas. On occasion, the financial market may react cautiously and negatively to the damage caused by risks. For instance, the stock market plummeted after the 2011 earthquake, tsunami, and nuclear facility calamities in Japan (Jones, 2015).

Since the 1970s, the effects of human activity on the environment have been increasingly recognized. International policy frameworks have been devised and implemented over the past three decades to mitigate and limit this effect. Nevertheless, given the current rate of dynamic global environmental change, human activities must be regulated to prevent and mitigate its effects. Scientists and policymakers have only recently realized the need for regulations that take into consideration the complexities of this interaction. The growing interest in climate change adaptation lends credence to this perspective. The scientific community emphasizes the importance of understanding both the fundamental factors that predispose individuals to peril and how environmental conditions can exacerbate such risks.

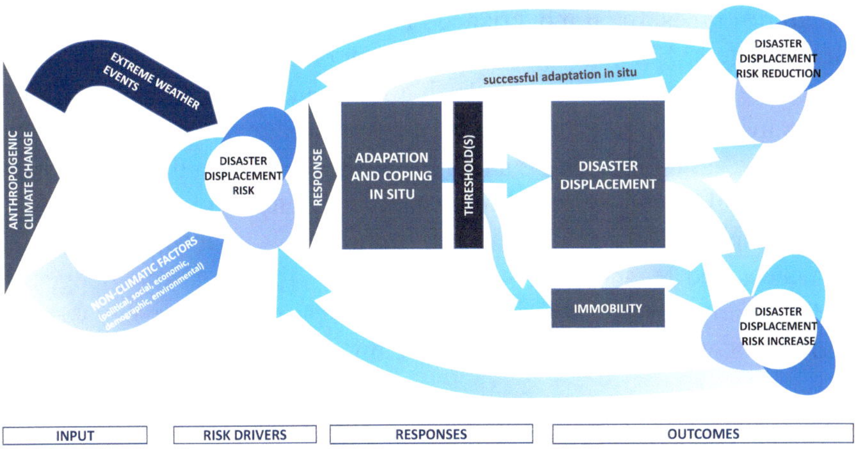

Figure 1.2: Relationships of hazards, vulnerability, and disasters.

Source: Thalheimer L, Heinrich D, Haustein K, Singh R. Integrating a Disaster Displacement Dimension in Climate Change Attribution. *Meteorology*. 2022; 1(4):468-476. https://doi.org/10.3390/meteorology1040029 [accessed 12 Jun, 2023]

Natural disasters do not occur at random. They result from a confluence of hazardous situations and dangers. In addition to revealing underlying social, economic, political, and environmental issues, disasters exacerbate them. These conditions impede growth by undermining achievements in infrastructure and technological innovation as well as in politics, society, and education. The Millennium Declaration urges all parties to "intensify our collective efforts to reduce the number and effects of natural hazards and man-made disasters," recognizing that natural and man-made disasters pose a threat to progress. According to a study, development investments are in jeopardy if disaster-prevention measures are not implemented.

However, few development organizations practice comprehensive project management, and fewer still are aware of how environmental management can mitigate disaster risk (Jones, 2015).

Human habitation patterns, livelihood options, and behavior harm human development, exacerbate environmental deterioration, and increase the likelihood of natural disasters. The impoverished are more susceptible to natural disasters because they are compelled to reside in the most remote locations and have the least access to disaster prevention, early warning, and preparedness. Due to their dearth of support systems, insurance, and other financial resources, the impoverished have the least ability to recover from natural disasters. A holistic approach to disaster risk reduction acknowledges the role of the environment in both the occurrence and prevention of disasters in communities. In addition, it acknowledges that natural disasters and the devastation they cause can have a direct effect on the environment. Environmental management canprospectively contribute to the reduction of catastrophic risk, including environmental science, information, governance, and technology. Importantly, this strategy acknowledges the need for environmental administrators, regardless of whether they work for the government or reside in rural areas (Paul, 2011).

1.2 HAZARDS AND CLIMATE CHANGE

Although the connection between the environment, development, and natural disasters is rarely questioned, the environment and multiple responsibilities have sparked considerable concern. Although it is widely acknowledged that natural disasters affect ecosystems, it is occasionally forgotten that protecting ecosystem services can help save lives and generate revenue.

1.3 NATURAL HAZARDS ARE PHYSICAL PROCESSES THAT CAN BE DIRECTLY AFFECTED BY DEVELOPMENT PROCESSES

Earthquakes, droughts, and flooding are deemed "dangerous" natural disasters because they pose a threat to human life or vital natural resources. Planetary physical processes manifest as threats. Contrary to conventional belief, human activities influence the timing, intensity, and frequency of tropical cyclones and droughts. Global warming caused by humans is increasing the frequency and severity of extreme weather events. Regional rainfall patterns are substantially impacted by deforestation, desertification, and the frequency of droughts (Jones, 2015). Climate change may exacerbate or alter the hydrometeorological hazards posed by heat waves, cyclones, floods, and droughts. Regional precipitation, temperatures, and weather patterns are influenced by human-caused emissions of greenhouse gases, which cause climate change. According to the Intergovernmental Panel on Climate Change (IPCC), climate change will strain vital ecosystems and cause food and water shortages in this century. Many regions of the globe may already be experiencing the effects of climate change. Experts believe climate change is increasing the frequency and severity of risks despite this warning. Even though reducing greenhouse gas emissions is one strategy for combating climate change, the scientific community believes that additional measures are required. People must adapt to the transition's consequences, which are already unavoidable due to past greenhouse gas emissions. According to the IPCC, combating present weather extremes may increase climate change resilience. Addressing current exposure to these extremes is the first step in preparing for climate change. Changes in regular climatic conditions and climatic unpredictability can affect a society's capacity to adapt to potential threats. The ability to adapt to climate change is contingent on several crucial calamity risk reduction strategies, such as early warning systems, risk assessments, and the use of sustainable natural resources.

1.4 HEALTHY ECOSYSTEMS PROVIDE NATURAL DEFENSES

According to the United Nations Millennium Environment Assessment, floods and fires have negative effects on the environment and human populations. In addition, it highlights the significance of ecosystems in reducing human risk. Ecological variables influence the magnitude and frequency of hazard occurrences, as well as the presence of natural barriers that may mitigate the effects of disasters and safeguard communities. Wetland ecosystems continuously consume and release surface water, precipitation, snowmelt, groundwater, and inundation (Paul, 2011).

The May 2004 debris flow killed an estimated 400 of 11,000 residents in Jimani, displaced up to 3,000 people, and destroyed at least 300 houses, accounting for 10 to 15% of the town's housing stock. The principal causes were the settlement's location on an alluvial fan, a big rainfall event, and deforestation in the upper watershed. Secondary factors included the area's geomorphology (poorly cemented sediments, gravels, and debris), watershed hydrology, and a lack of local competency for weather forecasting, river monitoring, and communications/evacuation orders. The most likely culprit for the sudden flooding calamity is a low-pressure system that passed across the Dominican Republic and Haiti from May 18 to May 25 and had its origins in Central America. This hurricane dumped more rain on the border regions of Haiti and the Dominican Republic than they normally get in a whole year. Jimani's location, which is built on an alluvial fan made of silt from flooding episodes that happened thousands of years ago, also had an important part in the disaster. The third major cause of the disaster was deforestation in the upper watershed. Over 80% of the catchment area of the Soliette River is located outside of Haitian boundaries, and it flows into the Dominican Republic. According to some estimations, the original forest cover on the Haitian side of the basin was "virtually treeless", with the majority of clearance taking place over the previous two decades (Jones, 2015).

1.5 DEGRADED ECOSYSTEMS REDUCE COMMUNITY RESILIENCE

Modern disaster managers have a better grasp of the variables that diminish the capacity of communities and social systems to endure disasters as a result of the change in emphasis from hazards to underlying vulnerabilities. Many of these features are influenced by their surroundings. Poverty, environmental degradation, and increased catastrophe risk are all linked. Because inhabitants are already trying to live, living in a peripheral or ecologically damaged region may make it more difficult for them to deal with additional stressors. Lack of access to healthcare and other services, competition for precious resources, ineffective governance systems, and restricted economic prospects can all impede a community's ability to respond to and recover from a disaster. Environmental management and community-based resource management can help communities become more resilient by supporting sustainable subsistence, minimizing conflict, and developing collaboration for good governance.

Even while the ecosystem usually recovers after a natural catastrophe, there is still a major risk to human life and livelihoods if environmental effects are not handled. As urban and industrial regions are built without respect for the threat posed by natural hazards, the potential for natural-hazard-caused crises grows. An environmental emergency occurs when a material is released into the environment uncontrollably, unexpectedly, or inadvertently. as well as unexpected recovery procedures that neglect ecosystem health and services Existing vulnerabilities may be repeated or increased during the recovery phase of urgent development if environmental considerations are not carefully considered. In today's world, there might be added dangers (Paul, 2011).

1.6 SOME ENVIRONMENTAL IMPACTS REQUIRE IMMEDIATE ATTENTION

A study of the environment and catastrophe would be incomplete if it did not acknowledge that environmental deterioration is a human-caused risk and hazard in and of itself. Local businesses and communities, for example, will continue to suffer as a result of desertification and biodiversity loss. The disaster reduction community's viewpoint on risk and susceptibility provides a valuable framework for studying patterns of sensitivity to environmental change and recommending ways to lessen such susceptibility.

1.7 THE HYOGO FRAMEWORK FOR ACTION

The Hyogo Framework for Action (HFA) was developed by countries at the World Conference on Disaster Reduction in Kobe, Japan, in January 2005. It serves as the framework for all global disaster risk reduction initiatives. The Hyogo Framework acknowledges that disasters are caused by a combination of hazards and environmental vulnerability and that environmental degradation increases the chance of catastrophes. Governments are urged to work toward "substantial reductions in disaster losses, lives lost, and the social, economic, and environmental assets of communities and countries." As a result, the Hyogo Framework recognizes the environment's sensitivity to hazards as well as its role as a catalyst for disaster risk. Civilizations' ecosystems have a direct influence on their evolution and progress. They interact to build a complicated system with varying levels of resistance and susceptibility. The less deteriorated a system's environmental component is, the more resilient and responsive the system as a whole (Paul,. 2011).

The Hyogo Framework includes "Reducing the Underlying Risk Factors" as a Priority for Action, and it expressly requests environmental and natural resource management in addition to other programs that: promote responsible ecosystem utilization and management, including improved land use planning and development practices to reduce risk and vulnerability; and implement disaster risk reduction strategies in environmental and natural resource management. Encourage the implementation of disaster risk reduction and climate change adaptation techniques that include the risks associated with present and future climatic variability and change. These methods should involve a distinct assessment of the disaster risks caused by climate change, the creation of specialized risk-reduction strategies, and better and more consistent use of climate risk data by planners, engineers, and other decision-makers (Jones, S. 2015).

The world was made aware of the need for governance in reversing environmental deterioration during the 2002 World Summit on Sustainable Development. Environmental managers with decades of experience mainstreaming an issue that crosses sectors and is hampered by a propensity to prioritize short-term gains above long-term security should be contacted. Institutions and strong governance are also necessary for catastrophe risk reduction. Effective governance necessitates a large number of efficient and responsive institutions, of which legislation is only one component. Regulations must be observed, and the political will is required to balance opposing interests and achieve necessary reforms must be emphasized. Furthermore, national governance mechanisms, such as multisectoral national forums, must be built and developed, as well as integrated with international frameworks such as conventions. To improve the way institutions, deal with environmental and disaster-related issues, environmental managers must be fully integrated into national disaster risk management systems, and risk reduction measures must be incorporated into environmental regulatory frameworks.

Risk data can help determine where, when, how, and why disasters are likely to occur. It manifests in a variety of forms, such as telemetric data that alerts quickly to seismic activity, climate projections that help explain long-term patterns, and storm forecasts. The risk information also includes explanations of susceptibility patterns and their underlying causes. Environmental conditions are monitored and observed using a variety of ground- and space-based equipment. Environmental data serve as the basis for spatial planning to establish the optimal buffer zones, land uses, and construction regulations. It functions as input for models, predictions, and projections that aid in risk prediction and reduction. Due to the dynamic nature of risk and vulnerability, they must be assessed continuously (Jones, 2015).

If individuals adopt a culture of prevention and are well-informed, disasters can be reduced significantly. Environmental and natural disaster-related traditional knowledge sources are typically maintained by local communities. Universities and other research institutions have developed additional data for policymakers and disaster management operations. It is still essential to construct an understanding of new technology risk reduction based on environmental principles. It is crucial to make available information about environmental catastrophes for use in fact-based investigations, political advocacy, operational decisions, and educational courses (Paul, 2011).

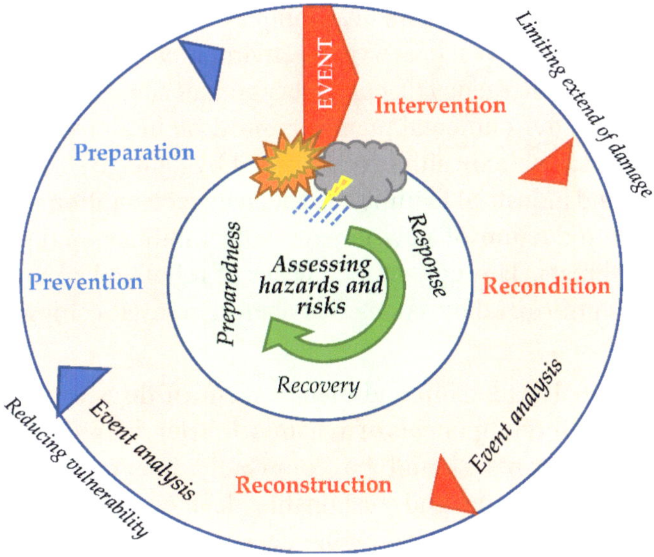

Figure 1.3: Disaster risk reduction.

Source: Matić BB, Karleuša B. Ecosystem-Based Disaster Risk Reduction Framework as a Tool for Improved River Basin Natural Water Retention Capacity and Environmental Hazard Resilience. *Environmental Sciences Proceedings*. 2022; 21(1):40. https://doi.org/10.3390/environsciproc2022021040 [accessed 12 Jun, 2023]

Risk is increased by changes in climate-related weather intensity, circulation, hydrology, and sea level. Changes in the severity of weather, circulation, hydrology, and sea level brought on by climate change increase risk. Large populations in densely populated coastal areas and flood basins are made more vulnerable by the loss of ecosystem services that control flooding and fires. To address the causes of unfavorable environmental conditions, it is necessary to develop governance structures, education, awareness, and capacity development mechanisms, as well as appropriate technologies founded on both contemporary science and traditional knowledge. Environmental management contributes to risk reduction by conserving and enhancing the natural conditions that facilitate resilience and climate adaptation.

For organizing evacuation, search and rescue operations, and the distribution of aid to survivors, it is essential to devise contingency plans in advance based on all available risk information. All required stakeholders must be involved in the development and implementation of effective disaster response plans to save lives, preserve the environment, and safeguard at-risk property: Effective emergency response necessitates communication between multiple parties. Environmental managers must be in a position to provide essential services. Public health is endangered by damage to environmental infrastructure and industrial facilities. Inadequate recuperation planning may result in the introduction of new threats, vulnerabilities, and security and sustainability threats. Throughout the phases of relief and recuperation, the environment is prioritized by saving lives before re-establishing livelihoods (Jones, 2015).

Strong national leadership and management of disaster risk reduction initiatives are usually components of dynamic disaster risk reduction systems. Environmental concerns should be completely integrated into national catastrophe risk reduction and sustainable development strategies, with environmental institutions acting as the basis for their extension. Another option is to include more environmental managers in disaster relief and recovery activities. Despite the environmental repercussions of catastrophe risk and susceptibility, as well as the long-term implications for sustainable development, environmental managers have only been involved in disaster prevention, response, and recovery when needed. Furthermore, disaster managers, who are required to participate in environmental initiatives, have given little thought to the environmental consequences of their jobs (Paul, 2011).

Environmental managers have effectively used a variety of regulatory frameworks to control future environmental deterioration by pushing developers to assess the possible environmental implications of their plans and activities. According to the International Federation of Red Cross and Red Crescent Societies, the ProVention Consortium is presently working on a guideline on the use of environmental impact assessment (EIA) methodology as a risk-reduction technique. Three critical EIA protocol stages are identified: A detailed impact assessment should be done for specific project categories in high-risk locations, and the environmental assessment approach should incorporate data gathering on natural hazard-related hazards as a critical early element of project scoping. It is also critical to perform a thorough assessment of natural dangers and their vulnerability.

Figure 1.4: Natural disaster risk management.

Source: Vučić N, Cetl V, Šantek D, Mađer M. Importance of Official Geodata in Disaster Risk Management—Case Study of Croatia. *Earth*. 2021; 2(4):943-959. https://doi.org/10.3390/earth2040055[accessed 12 Jun, 2023]

Other approaches, in addition to EIAs, are used in national environmental assessments. The most noteworthy example is the application of strategic environmental analysis (SEA), a tool that gives critical inputs for particular policies or programs in at least 14 Asian, Latin American, African, and Central European countries. The Sri Lankan government, for example, has sought to institutionalize SEAs to assess the environmental consequences and development choices offered in municipal planning papers, as well as the feasibility of adding disaster risk factors into SEA procedures (Jones,. 2015).

All development plans, risk reduction activities, and preparation planning should be informed by risk assessments. Environmental change must be addressed as a risk parameter in these national and local risk assessments. Rising sea levels and their influence on danger zones must be included in coastal risk assessment. The emphasis on environmental change as a risk factor reminds us that the notion of "risk" is variable and ever-changing. As a result, risk assessments must integrate dynamic risk through enhanced models and regular updates. Hazard risk mapping is an indispensable requirement for environmental management agencies with established mapping skills and tangible data storage foundations. Sadly, couples rarely share these maps. Local risk and vulnerability mapping must incorporate both physical and socioeconomic variables. In addition to identifying ecologically sensitive areas that could be devastated by a natural disaster, several environmental agencies have analyzed the threat of hazards (from inundation, landslides, seismic activity, etc.). The KenSea Environmental Sensitivity Atlas for the Coastal Zone provides data on the distribution and dispersion of natural resources along the coast, enabling the Kenyan command center to make tactical and strategic decisions in the event of an oil leak (Paul, 2011).

Indigenous people, the majority of whom reside in disaster-prone areas, rely on environmental protection and disaster management for their survival. Due to their years of experience and close connection to the environment, they have a multitude of knowledge regarding environmental dangers and occurrences. This information is crucial for emergency preparedness and environmental preservation in these regions. Given the loss of traditional means of subsistence and altering settlement patterns, it is difficult to maintain the continuity of traditional knowledge through generational transmission. When developing community-based calamity risk management strategies and actions, indigenous knowledge should be employed alongside scientific evidence.

According to the Hyogo Framework for Action, science, and technology are crucial pillars of disaster risk reduction. Even though neither national nor international organizations have taken a position on the issue, there is currently no scientific evidence linking environmental management to the severity and frequency of natural disasters. Although there are centers of excellence that can help us better understand these causal connections, increasing the participation of policymakers and operational management may encourage research that demonstrates cutting-edge techniques for mitigating catastrophic risk. Finding technical solutions and supporting environmentally friendly risk reduction innovation frequently requires additional funding (Jones, 2015).

Chapter

2 A BRIEF INTRODUCTION TO BIOLOGICAL HAZARDS, RISKS, AND DISASTERS

CONTENTS

2.1 INTRODUCTION

Biochemical substances, often known as biological hazards, endanger the health of living organisms. Source of Biohazardous include humans, bacteria, viruses, insects, plants, birds, animals, and plants. Some biohazard-caused diseases or infections, such as respiratory anthrax, can be easily passed from one person to another once they have spread. Biohazardous chemically-caused illnesses may need the deployment of suitable disease-prevention measures, depending on the severity of the disease and the population affected. The Centers for Disease Control and Prevention (CDC) classifies injuries based on their level of danger, with level 1 injuries posing the lowest risk and level 4 injuries offering the highest risk (DeWit, 2020).

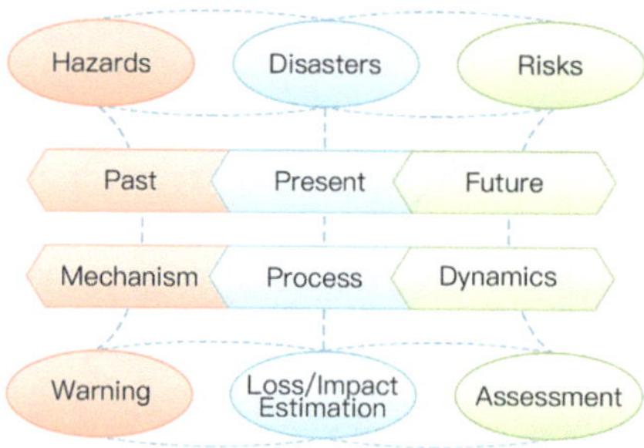

Figure 2.1: Relationships between hazards, disasters, and risks.

Source: Shi, P., Ye, T., Wang, Y. et al. Disaster Risk Science: A Geographical Perspective and a Research Framework. *Int J Disaster Risk Sci* 11, 426–440 (2020). https://doi.org/10.1007/s13753-020-00296-5

Biohazard category 1 agents do not cause disease, infect healthy humans or animals, or constitute a serious environmental concern. Non-pathogenic E. coli and Bacillus subtilis isolates are examples. Category 2 biohazards have little environmental effect and are related to curable or avoidable human illnesses. Lyme disease and the hepatitis A, B, and C viruses are two examples. Level 3 biohazards are chemicals linked to deadly or life-threatening human illnesses for which there exist therapeutic or prophylactic actions. Even though bacteria and viruses are seldom passed from person to person, they can cause severe to deadly sickness in both people and animals. Infectious illnesses such as hantaviruses and anthrax exist. Category 4 biohazards can infect humans and animals and produce serious or deadly diseases, but they pose no preventative or therapeutic concerns. They are a danger to the community and area. Individuals can easily transfer these diseases to one another, whether purposefully or accidentally. Hemorrhagic illnesses in Bolivia and Argentina include the Ebola virus and hemorrhagic fevers. When confronted with such serious biological threats, it is critical to wear a positive-pressure suit with a dedicated oxygen supply. There are currently no microorganisms categorized at this level (Davies, 2015).

When people are accidentally or intentionally exposed to biological substances, they may develop clinical or medical symptoms that act as early warning indications of the accidental or purposeful introduction. Environmental monitoring systems, such as BioWatch, can identify the presence of a biological agent in the environment and begin targeted environmental sampling, as well as thorough clinical surveillance, to rule out or confirm an occurrence.

All biohazard samples are transferred to a Laboratory Response Network (LRN) laboratory for confirmation testing due to the lack of acceptable field tests for biological agents. A network of public health laboratories is activated when a local health monitoring system or network detects an outbreak or epidemic with an unknown cause. The LRN is a network of laboratories at the municipal, state, and federal levels that can respond to bioterrorism and other public health emergencies. The LRN-Biological (LRN-B) laboratories can be reached through the LHD Communicable Disease Control Officer or the CDPH Duty Officer. There are several LRN-B laboratories in California, including the state laboratory (CDPH). LRN National Laboratories include the Naval Medical Research Center, the United States Army Medical Research Institute of Infectious Diseases, and the Centers for Disease Control and Prevention. Vaccines and other treatments for the highly contagious biological agents being investigated at the LRN National Laboratories are frequently in short supply. They also supervise, evaluate, and carry out bio forensics, agent activity investigations, and expert strain characterizations. Local agencies such as the Health Department (LHD), Environmental Health Department (EHD), police enforcement, fire/hazardous materials, and local emergency management are the major responders to biohazard occurrences, according to municipal legislation and practices. Activities should focus on controlling, regulating, and eliminating the illness, in addition to safeguarding, prioritizing, and caring for people, the environment, and property. Public health professionals, hazardous materials response teams, police enforcement, and emergency medical treatment providers may be required. An incident command or a unified command may be constituted according to the type, extent, and scope of the biohazard (DeWit, A. 2020).

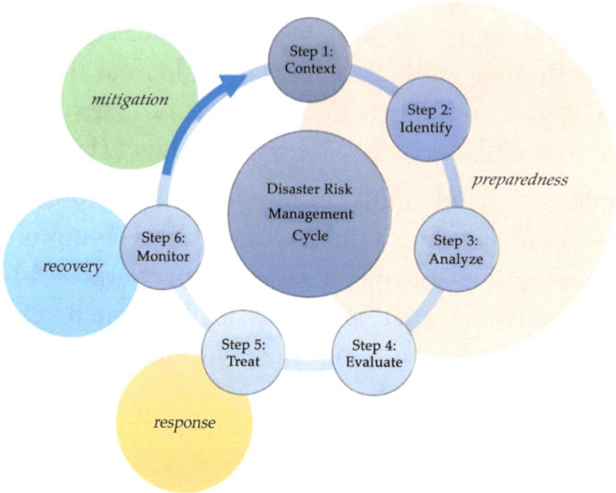

Figure 2.2: The disaster risk management cycle.

Source: Assilzadeh H, Levy JK, Wang X. Landslide Catastrophes and Disaster Risk Reduction: A GIS Framework for Landslide Prevention and Management. *Remote Sensing*. 2010; 2(9):2259-2273. https://doi.org/10.3390/rs2092259

Agriculture, cattle, and mining are the three most important economic sectors that are influenced by geography, season, and disease kind. Animals are more vulnerable to zoonotic viral outbreaks than people, and government travel restrictions and lockdowns may make it impossible for agricultural employees to earn a living. Another hurdle is the supply chain, which connects agricultural goods to markets and other businesses. A bug infestation or the consequences of a severe weather event on agricultural and agro-based companies might jeopardize the food security of millions of people. Absences from work, transportation congestion, a shortage of equipment, and closed markets are all possible issues for the mining business (Davies, 2015).

2.2 IMPACT ON THE SECONDARY SECTOR

Like the primary sector, the secondary sector will be affected by the geographic location, timing, and characteristics of the epidemic. Absenteeism, insufficient transportation, labor, raw material, and equipment shortages, as well as service interruptions, could all be detrimental to the manufacturing company. Because this industry is so dependent on vital infrastructure, any disruption to such infrastructure could have a significant effect on the manufacturing sector. In addition, public risk aversion and political constraints may reduce the demand for manufactured products. Nonetheless, there is a good chance that businesses that provide personal protective equipment, medications, and other necessities will benefit from the unanticipated increase in demand for specific items.

2.3 IMPACT ON THE TERTIARY SECTOR

When it comes to investing in new personnel, resources, and other criteria to ensure operational readiness, the healthcare industry in developed, emerging, and underdeveloped nations face significant obstacles. Additionally, hospitals are experiencing a decline in average patient visits, which strains their regular revenue stream. Transportation, tourism, and hospitality are interdependent industries that would be severely impacted by government travel restrictions and risk-averse consumer behavior. The media and information technology industries may be negatively affected by the emergence of fake news and cybercrime (DeWit, 2020).

2.4 IMPACT ON THE QUATERNARY SECTOR

The education sector is among the most frequently affected sectors. Principally responsible are teacher apathy and school closures. Due to children's susceptibility to numerous viral infections, school closures in the past have disrupted the academic schedule. This may result in developmental, psychosocial, physical, and emotional difficulties in the future. In addition, children may be particularly vulnerable to abuse, neglect, and other child safety issues during a health emergency. Due to worker emigration or absenteeism, a health emergency may also impede the operation of public utility services, impacting the operation of other sectors that rely on essential services. Moreover, the effect and sector response would vary according to the specifics of the biological hazards (Davies, 2015).

2.5 RISK ASSESSMENT AND SCENARIO PLANNING

To better comprehend the dynamics of biological hazards, it is essential to evaluate the risk variables, as well as their reciprocal and interactive relationships. When intending to conduct a risk assessment of biological hazards, the following essential factors must be considered.

2.5.1 Identifying Biological Hazards

Risk assessments serve as the basis for planning and execution by analyzing potential hazards and evaluating current exposure and vulnerability conditions that have the potential to harm people, property, services, and the environment. This strategy involves the examination of both historical and current risk factors, as well as anticipated future development threats. To inform policy decisions, risk assessment activities must be comprehensive and cross-sectoral to encompass the entire continuum of risks, including those posed by biological hazards.

2.5.2 Categorizing Biological Hazards for Risk Assessments

Based on what is known about them, biological hazards can be classified as either existent or emergent for risk assessment and management. Even though all biological hazards have the potential to cause a wide range of emergencies of varying types, durations, and degrees, they cannot be classified as emergency or non-emergency based only on their propensity to cause epidemics, pandemics, or PHEIC. Because there is so much information available on the causes, mode of transmission, treatment options, potential impact, etc., risk assessment of current biological hazards, particularly those with the potential to cause an emergency (such as malaria, cholera, etc.), can be completed. A strategic risk assessment can be conducted to allow proactive preventive, preparedness, public health, capacity development, and monitoring planning. Emerging biological threats, on the other hand, are erratic, imprecise, and dynamic, making this evaluation difficult (DeWit, 2020).

According to the 2017 National Disaster Risk Assessment (Biological Hazards Risk Assessments), risk assessments for health threats at the intersection of animals, people, and ecosystems are required for approximately 75% of newly emerging infectious diseases in humans that have emerged as a result of zoonotic pathogen exposure (Davies, 2015).

Emerging biological hazards are dynamic occurrences that rely on a variety of factors to accelerate or restrict their spread due to their ambiguity, unpredictability, and inclination to reoccur as new waves (which frequently operate differently from the past waves). A fast, dynamic risk assessment is necessary to comprehend the spatial and temporal dimensions, causes, and numerous consequences of the risk on various businesses and social groups. To improve the agility of risk-informed decision-making, a fast risk assessment will aim to capture the dynamic conditions and, as a result, the changing degree of risk for an individual, a community, and multiple sectors at various phases of the hazard's transmission. Examining novel biological hazards is crucial for demonstrating this dynamic nature since they influence other risk assessment criteria such as potential exposure, vulnerable sectors and populations, and knowledge gaps. A diverse team of epidemiologists, virologists, biotechnologists, health professionals, decision analysts, DRR experts, and others must provide data on critical aspects such as the mode of transmission (airborne, waterborne, surface transmission, soil-borne, human-to-animal, etc.), the incubation period, the rate of transmission, recovery, and mortality, and, if applicable, any stages of transmission (such as the onset of disseminated disease).

2.5.3 Linkages with Integrated Surveillance Systems for Capturing the Dynamic Nature

A newly emergent biological danger may now be detected by integrated surveillance systems, allowing for a swift and dynamic assessment of the threat. This system should include hospital monitoring, laboratory monitoring, veterinarian monitoring, community monitoring, and external monitoring in addition to sickness surveillance.

2.5.4 Assessing Direct Exposure of Transmission of Biological Hazards

Direct biological hazard exposure happens often as a result of their dissemination throughout a territory, and it may be predicted based on the variables driving their spread (as identified during hazard assessment). This varies by risk since many biological risks have diverse transmission channels and affect different social groups (depending on age, comorbidity, pregnancy, links to specific vocations etc.). The identification of groups at risk of transmission is dependent on a variety of factors, including the exposed population's immunity, the dosage, and the length of exposure. The duration and severity of direct exposure might vary based on the stages of biological hazard transmission, necessitating scenario planning and modeling to anticipate or extrapolate the exposure (DeWit, 2020).

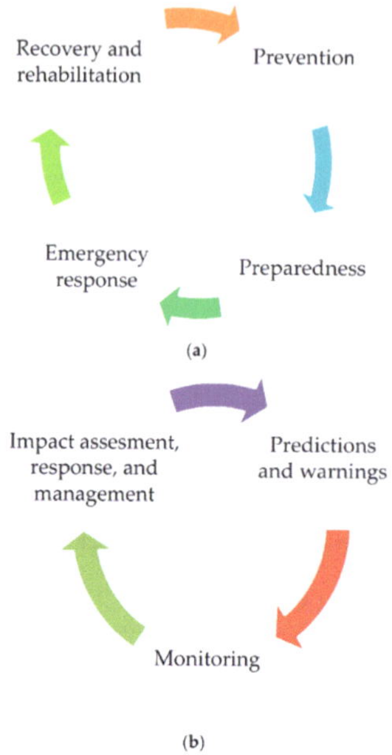

Figure 1.5: The integrated risk management cycle.

Source: Yang T-H, Liu W-C. A General Overview of the Risk-Reduction Strategies for Floods and Droughts. *Sustainability*. 2020; 12(7):2687. https://doi.org/10.3390/su12072687 [accessed 12 Jun, 2023]

2.5.5 Assessing Multi- and Cross-sectoral Dimensions of Exposure of Biological Hazards

While the term "direct exposure" is commonly used to characterize the transmission of a biological hazard, it is not always synonymous with "indirect exposure." The techniques used to prevent and manage the spread of biological hazards will affect how much organizations and individuals are exposed indirectly. A strategy including the region's total shutdown, for example, would have a drastically different impact on the exposed persons and industries than a strategy featuring a partial lockout and the continuing of key service delivery. Officials can anticipate areas, populations, and sectors that are at risk of exposure by conducting scenario planning exercises that test potential strategies for the various stages of transmission and subsequent waves of transmission (including the effects of exposure on urban, geographically isolated, and disadvantaged areas). Based on the findings of these exercises and the risk assessment suggestions, multisectoral inter-

and trans-disciplinary teams should be able to prioritize geographic areas, social groups, sectors, infrastructure, and services under short-, medium-, and long-term scenarios (Davies, 2015).

2.5.6 Assessing Vulnerability to Biological Hazards and Systemic Risk

For the examination of exposed persons, sectors, and systems' vulnerability to biological hazards, it is critical to consider their multiple underlying features, some of which may differ from one biological hazard to the next. As a result, unlike other hazards that frequently identify and know vulnerable groups and sectors in advance (such as older people, people with disabilities, children, women, marginalized minorities, LGBTIQs etc.), biological hazards can have varying effects on different groups, making it difficult to identify vulnerable groups in advance. These groups and sectors are subject to change as a result of the control and containment mechanisms used, as well as the transmission stages. As a result, it is critical to assess these categories' possible short-, medium-, and long-term implications (DeWit, 2020).

2.5.7 Evaluating Existing Capacities for a Prompt Response and Early Recovery

Examples of resources and plans that should be identified and evaluated as part of capacity assessments include specialized hospitals, diagnostic facilities, testing facilities, treatment capacity, availability of personal protective equipment, trained and equipped essential service providers, existing protocols, public awareness, behavioral factors, etc. It is essential to examine these resources critically to identify any voids or obstacles that may hinder the execution of an effective and efficient response during disaster management, especially those associated with the production and distribution of necessities such as PPEs, ventilators, etc. Using scenario planning (corresponding to distinct transmission phases), the efficacy of the resources required for various transmission stages and future transmission cycles can be assessed. The ability of critical and essential services to function with limited human and technical resources and partially disrupted supply chains; current and anticipated production and supply of essential materials; sectors' ability to adapt to alternative or virtual functioning; capacity of ICT infrastructure to support virtual operations in remote and rural areas; and capacity of local communities (socioeconomic and psychological) should all be considered. Techniques such as scenario planning and network modeling can be utilized to comprehend and evaluate the potential interdependencies between biological risks and other natural or man-made hazards. A biological hazard-induced disaster that occurs sequentially or simultaneously with a natural disaster. Each scenario presents a unique level of danger. It increases the risk to emergency responders and afflicted individuals in relief centers due to the potential for the transmission of biological hazards in a crowded environment or a flooded location (in the event of a water-borne illness pandemic). Considering the possibility of cascading and sequential/simultaneous disasters while evaluating all risk factors, such as hazard, exposure, susceptibility, and coping capacity, is required. This will allow for the implementation of risk-informed response planning and management (Davies, 2015).

2.5.8 Understanding the Interplay of Multi-Hazard Risks Through Scenario Planning and Network Modeling

The foundation for biohazard risk assessments is the availability of reliable and accurate data, information, expertise, and knowledge in a variety of sectors. These skills and knowledge must also be shared between industries and nations. Given the obvious effects that biological hazard-induced disasters may have on the regular operation of society and the performance of nearly every sector of the economy, it is essential to incorporate them into sectoral and overall development planning in addition to DRR planning.

2.6 PREVENTION AND PREPAREDNESS

Consequently, each sector must examine its operations through the prism of disaster risk management, identifying underlying vulnerabilities, exposing infrastructure and supply chains to risk, and working toward risk prevention, mitigation, management, and overall resilience. In contrast, planning alone is insufficient to respond to and recover from biological hazards.

2.6.1 Understanding the Non-Linear Transition of Biological Hazards from One Phase to Another

Responses to and recovery from biological threats, particularly pandemics, are not always linear. Instead, they keep reappearing as fresh surges with noticeably different impacts. With the arrival of each new surge, the population returns to the reaction phase until immunity or a vaccine is created. This should be recognized and included in sectoral planning and response strategies, as well as the possibility of employing the initial wave of occurrence (even in the case of unique hazards) as a potential early warning for following waves. Then, proactive actions to prevent, minimize, and prepare for future waves can be adopted (DeWit, 2020).

2.6.2 Institutionalization and Harmonization of Database Management

Biological hazards commonly impact different demographic groups differently depending on their intrinsic features (e.g., the elderly, pregnant women, people with co-occurring illnesses, etc.). For eliminating the sources of biological hazards, accurate and trustworthy baseline data, such as geographic and multitemporal data, data on exposure, vulnerability, and capacity, data on pre-disaster health, and population socioeconomic indicators, are critical. Data and information standardization and interoperability within and between nations and regional partners can considerably improve the efficacy and applicability of DRR initiatives. Furthermore, during an epidemic, these databases may be used to plan targeted interventions and early recovery, as well as conduct fast risk assessments and reach out to the most susceptible. They can also affect future biological hazard risk management by increasing understanding of the underlying reasons that lead to vulnerabilities. Interoperable, institutionalized sector-specific database management enables the development of sectoral strategies that are effective in mitigating the short-, medium-, and long-term effects on societal groups such as migrant populations, daily wage earners, people with disabilities, the poor, orphans, and SMEs, among others.

2.6.3 Making the Best of the Early Warning and Early Detection of Biological Hazards

Nations are frequently unprepared to deal with crises caused by growing biological threats. People tend to underestimate the risk of a disease spreading within their nation, even if it has been found elsewhere in the world. This is typically a squandered chance for early identification and prevention of transmission. Strengthening bilateral, regional, and global intergovernmental forums for proactive transboundary information exchange and active collaboration (research, reaction, finance, border management etc.) will help to detect and mitigate biological threats earlier. This collaboration is also required to prevent the biological threat from spreading too far. Similarly, each detection of a subnational unit should work as an early warning system, triggering emergency standard operating procedures (SOPs) across the country.

2.6.4 Resilient Health Service Delivery Systems

Given the importance of the health sector in avoiding and responding to biological threats, developing a strong healthcare delivery system should be one of the top goals. Primary/community health care agencies, as well as public and commercial health service providers, are included. One of the system's critical components is the infrastructure for delivering health services, which consists of a network of screening and testing facilities and laboratories, isolation and quarantine units, clinics, specialized hospitals and treatment facilities, and post-recovery monitoring facilities. Other components include a competent professional-staffed medical emergency response/surge capacity (Davies, 2015).

2.6.5 Dynamic Emergency SOPs for Multi- and Cross-Sectoral Intervention

Certain biological risks, in contrast to other hazards, have a long lifespan and can thus have a direct or indirect influence on almost all sectors of society, albeit to varied degrees. When faced with biological threats, governments may seek cooperation from relevant industries. For example, they should approach the manufacturing sector for help in addressing the increased need for PPEs and critical medical supplies. To guarantee a speedy reaction and recovery, a "beyond the health sector" approach to preparedness is required. Standard operating procedures (SOPs) for regulating biological risks should be designed collaboratively to meet varied requirements and implications. SOPs for current biological hazards may be produced using prior experiences and lessons gained, whereas SOPs for emerging biological hazards can be developed using worst-case scenario planning, such as the occurrence of cascading, sequential, and simultaneous biological hazards (DeWit, 2020).

2.6.6 Risk-Informed Sectoral Planning and Development

Using scenario planning to undertake a multi-sectoral, multi-hazard risk assessment may be immensely beneficial for bringing people from many sectors together to engage in risk-informed sectoral planning and development. Such planning would aid in the establishment of vertical (intra- and inter-sectoral) and horizontal (inter-sectoral) links in the framework of catastrophe risk reduction and sustainable development. Before a disaster, multi- and cross-sectoral exposure and vulnerability assessments can help decision-makers on numerous levels by raising awareness of regional circumstances and their consequences for the overall national risk assessment.

Land use and land cover planning, baseline information, and development at all scales will assist to minimize the underlying vulnerabilities and exposures of each sector and the general public under the guidance of the risk assessment. This helps to reduce the likelihood of present threats, address current systemic concerns, and prevent and mitigate situations that might lead to biological hazards and their spread.

Improvements in urban planning to prevent settlement overcrowding; strengthening of rural infrastructure; workforce skill development; prevention of occupational hazards; incorporation of a culture of preparedness and safety in sectors; and development of disaster-resistant infrastructure, services, and supply chains are all examples of risk reduction measures. To prepare for residual risk management, each sector must do sectoral contingency and continuity planning, establish methods of adapting to alternate modes of operation/service delivery/business, and build capacity at the lowest administrative level. Alternative solutions should take into account employee welfare, labor and safety legislation, and other relevant restrictions (Davies, 2015).

2.6.7 Strengthening Social Infrastructure and Socio-Economic Protection

Robust social infrastructure and services, such as those devoted to health, public health, nutrition, education, social and financial protection, etc., are necessary to reduce the susceptibility of vulnerable individuals and enable the successful management of biological hazard catastrophes. If comprehensive and appropriate socio-economic protection is not provided, these disasters will exacerbate existing social and economic vulnerabilities. By investing in social infrastructure and expanding the scope and effectiveness of socioeconomic protection systems to include all vulnerable groups, their susceptibility to biological and other risks can be reduced. These treatments should be driven by baseline data and socioeconomic risk evaluations. In addition to providing social security for the excluded, it is essential to establish informal support networks, which in many countries serve as the primary coping mechanism (DeWit, 2020).

2.6.8 Efficient Use of ICT and Innovation

Through the innovative application of information and communications technology (ICT), including new and emerging technologies that are contextually appropriate for local needs and challenges, it is possible to manage the unknown and unpredictability that distinguishes biological hazards from other dangers. ICT has immense potential to aid in the management of various biological hazard-related aspects, including prevention, mitigation, emergency warnings, early detection, risk communication, safety measure dissemination, active surveillance, exposure tracing, rapid testing and treatment, determining an individual's exposure and risk, and providing essential services. ICT also enables the remote operation of government organizations, services, and businesses. Several examples include tele-medication (for non-emergency patients or patients in remote locations), remote learning, teleworking, and company continuity during lockdowns and mobility restrictions. In addition, despite physical distance and limited mobility, ICT-based media may provide a means for individuals to maintain social ties. It is necessary to expand ICT infrastructure to ensure that no region falls behind, especially in remote and underdeveloped areas. As structural obstacles impede disadvantaged groups from having equitable and accessible access to ICT infrastructure and services, it is necessary to address gender equality, digital literacy, and socioeconomic disparities. Increasing reliance on digital infrastructure necessitates addressing cyber security and data protection issues (Davies, 2015).

2.6.9 Effective Risk Communication for Preparing Communities and Mobilizing Cooperation

The general population is significantly more involved in disasters resulting from biological hazards than in disasters resulting from other causes. In actuality, public participation is required to "break the chain of transmission." To emphasize the need for a "whole-of-society" approach to DRR, more individuals must be educated and made aware of the issue. Because risk communication is a two-way street, it is crucial to comprehend and compassionately address the perspectives, knowledge, attitudes, and socioeconomic and cultural practices of all societal segments. Effective risk communication requires communicating with communities in their native languages to ensure that minorities, particularly indigenous people, immigrants, and refugees, are not excluded. Government, civil society, non-governmental organizations, and the business sector should collaborate to communicate with their employees, partners, and the general public. Local stakeholders, particularly community and religious leaders, must play an active role in enhancing local governance in this regard. Mask use, proper coughing technique, excellent personal hygiene, proper PPE disposal, collaboration with critical service providers, avoiding frantic buying and stockpiling, and avoiding stigmatization of those infected or exposed or their professions should all be standard procedures. Risk communication is essential before a calamity to inform the public of the potential repercussions of their actions, thereby preventing the formation of new hazards and reducing the severity of existing ones (DeWit, 2020).

Chapter

3

DECLINE OF BEES AND OTHER POLLINATORS

CONTENTS

3.1. INTRODUCTION: THE BEE

The alarming rise in climate change, pesticide use, habitat loss and colony collapse put bees in danger. Bee populations have been steadily declining at the national, regional, and local levels. According to studies, various bee species have been steadily disappearing since the 1900s. As captured in the list developed by the International Union of Conservation (IUC), more than 18 bee species are in danger of being wiped out. The IUCN further mentions that more than 150 bee species are either vulnerable or extinct. In 2017, the rusty-patch bumblebee became the original bumblebee species to be called endangered. In 2016 alone, seven species of Hawaiian yellow-faced bees were recognized as endangered (Lawrence, 2015).

The Center for Biological Diversity (CBD) is submitting a petition to include the American bumblebee on the endangered species list. Similar to the rusty-patched bumblebee, experts state in a September 2021 report that the American Bumblebee was "previously a wide-ranging species." However, this is no longer the situation, and protection is essential. As per the report, the American bumblebee population has "dropped by 89% in relative numbers and continues declining toward eradication." This is because many causes, including loss of habitat, pesticides, climate change, and infection continue affecting the population. Over the previous two decades, it has nearly vanished from eight states and become rarer in many others (Altieri, 2013).

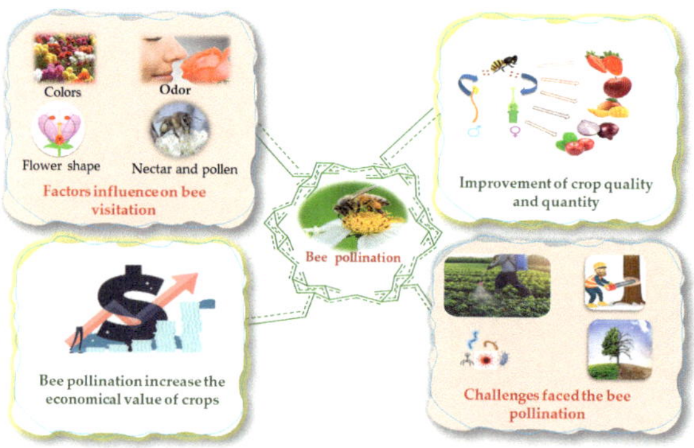

Figure 3.1: The effectiveness of wild bee pollination.

Source: Khalifa SAM, Elshafiey EH, Shetaia AA, El-Wahed AAA, Algethami AF, Musharraf SG, AlAjmi MF, Zhao C, Masry SHD, Abdel-Daim MM, et al. Overview of Bee Pollination and Its Economic Value for Crop Production. *Insects*. 2021; 12(8):688. https://doi.org/10.3390/insects12080688

Numerous factors, including harmful pesticides that deplete bee populations and habitat loss, pose a threat to bee species. Chemicals that harm bees can paralyze or eradicate them. But what is the primary threat to bee species? Climate change. We are witnessing a shift in everyday weather patterns due to climate change. This means severe weather, like extreme storms, heat waves, and winters without snow are becoming common. Additionally, although some climate change can be caused by natural factors, the changes we are facing are indeed caused by global warming—the warming of the Earth's atmosphere. Bees are unable to reproduce if they lose their habitats to deforestation or severe natural disasters. They won't be able to pollinate as a result. Bees may also be harmed by the following: pollution, colony collapse disorder, pesticides, disease and pests, loss of habitat, and climate change.

- ***Climate Change***: The various bee species are affected to different degrees by changes in weather and temperature. Bees and the products they pollinate are also affected by climate change, just as humans are. Due to their low heat resistance, large bees such as carpenter bees, eastern honeybees, and bumblebees, are more affected by rising temperatures. The decrease in the huge types of honey bees, particularly bumblebees, diminishes long-distance pollination (Lawrence, 2015).
- ***Loss of Habitat***: Construction projects by humans disrupt the bees' natural habitat. Street and building developments, mineral exploitation, and human settlements have resulted in habitat destruction and fragmentation. The *Colletes* and *Andrena* species of ground-nesting bees require loose soil to construct their nests. Nevertheless, they're at risk for extinction due to widespread industrialization that makes their home environments break down. Intensified land use divides bees' natural habitats, creating a movement barrier. Bees' health, range, and pollination activities all suffer as a result of this isolation (Altieri, 2013).
- ***Pesticides***: These are chemicals that kill bees. Honey bees, in particular, suffer from memory and learning impairment as a result of the toxic chemicals in pesticides. Additionally, it affects the bees' reproductive capabilities, decreasing the number of offspring produced. Pesticide compounds are called Neonicotinoids; multiple studies have shown that they can affect bees' health.

- ***Microorganisms and Pests***: Microorganisms are infectious bacterial conditions that impact bees and their production levels. Common pathogens that kill bees include:
  - Chalkbrood
  - European Foulbrood
  - Nosema
  - American Foulbrood and
  - Stonebrood.

  Bee larvae are frequently affected and killed by brood pathogens. A number of these infections cause impairments, such as wing deformation and cell decomposition. Bee colonies are disrupted partly by invasive species such as skunks, mice, and ants. The most harmful insect to bee health is the *Varroa mite*. A whole bee colony can be decimated by these mites. Moreover, skunks and mice attack hives, destroy bees and consume their combs (Lawrence, 2015).
- ***Nutritional Deficiencies***: Nutrition is critical to the population of bees. Pollen provides the majority of their required nutrients. Adult bees require carbohydrates, whereas their larvae need, fats, protein, and vitamins for healthy development. Aside from the aforementioned nutrients, bee growth is aided by essential compounds like amino acids. Nevertheless, not all pollen contains the appropriate amount of amino acids for larval health. There are various bee species and feeding habits. Some bees are drawn to particular kinds of flowers, whereas others pollinate particular flowers. For instance, the squash bees only consume pollen from gourd and squash plants. Honey bees and bumble bees, among other species, consume pollen from any flower.
- The endangered bee species also suffer from malnutrition as a result of habitat loss and pesticide use. For example, the *Apis mellifera* population of honey bees is among the most endangered species because they suffer from nutritional stress caused by pathogens, pesticides, and habitat loss. Escalated farming activities likewise cause the exhaustion of nutrients among bee species. It disturbs the natural range of crops and flowers that the bees need to survive (Altieri, 2013).
- ***Pollution***: Many researchers investigated how air pollution

affected bee populations. The atmosphere contains fuel exhaust and other toxic chemicals. These toxic materials are harmful to human health. Nevertheless, research into the effects of air pollution on pollinators, particularly honey bee populations, is still ongoing. Bees' bodies are covered in hairs that hold microscopic particles. These tiny particles come from flowers, though other things may also get caught between the hairs. Researchers discovered that bees frequently interact with plastic materials, and honeybees are frequently the most prevalent carriers of microplastics.

Additionally, flower scents are used by bees and other pollinators to find pollen. Pollution of the air filters out flower scents, which reduces pollination. Researchers in India investigated the decline of the *Apis dorsata*, a giant honeybee living in South and South East Asia. It is accountable for India's honey production. Researchers discovered that bees avoided polluted areas by foraging from plants and that more bees were found in less polluted areas. In other studies, 80% of bees from polluted regions perished, while 80% of bees found in rural and low-pollution regions thrived.

- ***Colony Collapse Disorder***: This is a condition in which adult worker bees rapidly decline. Colony collapse disorder upsets managed bee colonies more frequently than wild bee colonies when there are no adult worker bees present. It became a major reason for concern when nearly 60 percent of winter colony losses in the US connect back to CCD. Additionally, the rate at which CCD affected honey bee colonies in Europe rose by more than 30%. Different conditions, like chronic bee paralysis (caused by the chronic bee paralysis virus or CPV), an infection that makes whole colonies to perish, are viewed as an evolving danger to honeybee numbers.

3.2 REASONS TO PROTECT BEES FROM EXTINCTION

Honeybees pollinate around 80% of the world's crops, but they are vanishing at a rapid rate. Plants that are dependent on bees for pollination would cease to exist. Different ecosystems will undergo rapid changes if bees and the florae they pollinate disappear from the planet. Bee-feeding animals and plants that require pollination will both lose their food supply as a result. The disappearance of bees will not only have an impact on the environment.

Our food supply will undergo significant changes. Our favorite fruits and vegetables, such as apples, melons, and broccoli, are pollinated by bees. There may be a shortage of these fruits and vegetables on supermarket shelves if bees aren't present (Lawrence, 2015).

Because the stability of human populations is largely dependent on the solidity of bee populations, the ecological factor of the possible extinction of bees is also a humanitarian issue. Taking into account the significant role of honey bees' fertilization activities in supporting our farming practices, their disappearance would probably bring about a food emergency for humankind. Because they require the assistance of pollinators to reproduce, supplies of foods like berries, avocados, apples, coffee, and onions would considerably decline. With deteriorating supplies, the costs of these food sources would soar, making them harder to access by many people. More biodiversity would be lost as a result of the secondary effects of these plants' extinction on animals other than humans that also rely on them for food. Bee and other pollinator populations must be actively protected and nurtured to help us achieve SDG 2 and meet the world's rising food demand (Altieri, 2013).

- ***Honey Has Restorative Purposes***: Honey, a natural sugar with numerous benefits, is a popular health product made by bees and should be protected because it is sustainable and healthy. It has amino acids, calcium, niacin, zinc, potassium, magnesium, and other micronutrients. Honey is likewise a good source of antioxidants and promotes immune system function. Bee hives can also produce beeswax, in addition to honey. Propolis is another healthy product derived from bees. It is used by bee colonies to seal cracks and crevices. Vaseline often has it added to it to help soothe burns. Additionally, propolis and salve are usually combined to treat wounds and cuts. Bee venom is acidic and comes from bee excretion, however, amino acids, enzymes, and sugar are among the anti-inflammatory substances in it. Likewise, it diminishes allergic bee reactions and is often used for restorative purposes.
- ***Bees Secure the Food Supply Chain***: The ecosystem has several different pollinators, but native bee populations have a greater impact on the food supply chain. They are crucial to the agriculture industry's survival. Bees provide a variety of food by pollinating different plant species. Animals, bee populations and humans all benefit from the pollination process. As bees travel from one plant

to another, they consume pollen dust, which offers them sufficient energy to move pollen across distances. Crop yield increases when bees fly for an extended duration of time. Pollination is beneficial to other animals as well as crops and fruits for human consumption. Bees pollinate nuts, fruits and seeds, which many animals also consume. A decline in bee populations would result in human and animal famine and starvation. To avoid these issues, it is essential to safeguard endangered bee species.

- ***They Help in Economic Growth***: Before determining how much bees are worth financially, there are various factors to take into account. Since they drive economies through agriculture, bee populations have contributed significantly to different economies. Forbes estimates that the value of pollination by bees and other pollinators is approximately \$500 billion. Among the different species available, honeybees are the most valuable. They're valued at \$20 billion in yearly income. The honey produced by honey bees is highly valuable. Honey bee species are valued at about \$4 billion yearly.
- ***Bee Species Help To Protect Plant Ecosystems***: These pollinators not only help the ecosystem function properly but also allow for an abundance of food supplies. Due to their inability to reproduce, various plants would expire if pollinators were absent. Bees are essential to the healthy growth of food and flowering plants. Different kinds of flowering plants are pollinated by different species of bees. Bees and plants are inseparably linked to one another. Plants will also face extinction if bees disappear (Altieri, 2013).

3.3 HOW TO SAVE THE BEES

In the same way that flowers and bees exchange benefits, humans also benefit from bee pollination and should therefore ensure that bee populations remain healthy.

- ***Stop and Reverse the Loss of Habitat:*** If there aren't enough sites for bees to set up residence and acquire food, their populations will continue to drop. . Forestalling deforestation can consequently prevent the extermination of honey bees, as an incidental effect, and promote higher harvest yields. The nectar from certain native flowers, which is essential for bees' survival and is found in a skyscraper-lined, highway-crossed, cement-paved urban environment, is far off between. Availing bee-attractive flowers concentrated within a small area, similar to a part of your yard, can be helpful. Honey bees can store food reserves for the rainy season and winter survival thanks to this dense forage.

Preferably, maintain different flower and tree species with a decent measure of flowers from different species. Since the varieties of ornamental flowers that are regularly grown in yards don't create sufficient nectar for bees, consider bee-attractive flowers such as sweet alyssum, sunflowers, crocus, coneflower, honey bee salve, poppies, butterfly weed, and geranium (Lawrence, 2015).

- ***Eat Organic***: Most of our food depends intensely on the use of pesticides, which can be harmful to honey bees. Bees will be exposed to these toxic chemicals less if pesticides are used less to produce food. Since organic produce is grown without pesticides, it does not contain bee-killing chemicals. The food is produced without toxic chemicals, and purchasing organic honey helps support natural hive bee populations.

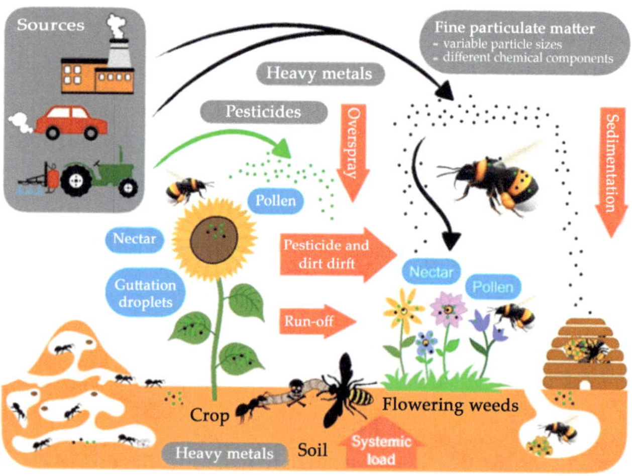

Figure 3.2: Effects of pesticides on bee species.

Source: Feldhaar H, Otti O. Pollutants and Their Interaction with Diseases of Social Hymenoptera. *Insects*. 2020; 11(3):153. https://doi.org/10.3390/insects11030153

The decline in beekeeping is a direct result of the social change brought about by increased urbanization. Individuals who reside in the city don't normally have the space to keep honey bees like those from rural areas. Therefore, It is critical to purchase local honey. Honey producers in the area conserve bee populations, but they need finances to operate their farms and look after themselves. Buying locally cannot only help the survival of small-scale farmers but also the plant and bee community biodiversity in an area.

- ***Limit the Use of Pesticides***: The European Food Safety Committee (EFSA) has expressed that three key pesticides present a "high acute risk" to bees and have been discovered at risky levels inside their hives. Pesticides that have been approved by governments must be tested to ensure that they do not harm pollinator populations, which, in the long term, will help improve crop yield and encourage sustainable farming. Before purchasing honey, check the label to see if it contains chemicals that are harmful to bees. Reduce the use of herbicides, which may not be directly harmful to honey bees, yet may still eradicate flowering "weeds" that bees require for food (Altieri, 2013).
- ***Beehives Shouldn't Be Destroyed***: Bee populations require any assistance they can get because there are so many things that are against their numbers. A few mating pairs within an area may mean the difference between a healthy population of bees and none at all. While many individuals are afraid of bees, it is essential to remember that, unless provoked, they are generally harmless. Not all bees sting – some species don't have the stinging capacity, and male bees lack the stinging capacity. Indeed, even female bees that truly do sting will only sting when provoked. You can get in touch with a local beekeeper association to have the bees relocated to a new area if you don't like having them around.

3.4 OTHER POLLINATORS

3.4.1 Birds

The conservation status of pollinating mammal and bird species is declining, according to recent research by IUCN and colleagues, with more species moving toward extinction than away from it. Approximately 2.4 bird and mammal pollinators each year have been relocated to the IUCN Red List category of extinction, representing a significant expansion in extinction risk across this range of species. According to UNEP's World Conservation Monitoring Centre, there is a stressing pattern that might be adversely influencing global pollination services, valued at roughly US$215 billion." It is known or inferred that pollinators exist in nine percent of all known bird and mammal species. Bats are the primary pollinators of many economically and ecologically significant plants, including agave and cacti, among other mammals. The White-eye, hummingbird, honey eater, and sunbird are important pollinators. Animals pollinate approximately 90% of flowering plants, and many of these species are heavily used by humans for food, medicine, materials, livestock forage, and other reasons.

According to Senior Scientist Michael Hoffmann of the Species Survival Commission (SSC) of the IUCN, "Most pollination is performed by invertebrates, such as bees, but regrettably minimal resources for species assessments means we cannot yet define the global status trend of these pollinator species." Nevertheless, these initial findings regarding bird and mammal pollinators are not encouraging for insect pollinator trends (Lawrence, 2015).

Habitat loss caused by unsustainable farming practices was viewed as the primary driver of decline for an extensive number of bird and mammal species. Hunting for bushmeat has a significant negative impact on pollinating mammals like large fruit bats, while invasive alien species have a negative impact on birds.

From 1988 to 2012, 18 species of pollinator birds met the criteria for "up-listing" to a higher-risk category. For instance, the Regent Honeyeater (*Xanthomyza phrygia*) was uplisted from Endangered to Critically Endangered as a result of rapid population decline caused by drought, habitat loss as a result of historic agricultural clearing, and possibly competition from other species. There were no pollinating bird species that were eligible for "down-listing" into lower threat categories.

Between 1996 and 2008, a total of 13 pollinating mammal species were up-listed to a higher risk level, while two species were relegated to a lower risk level. Among non-flying mammals, the Sunda Slow Loris (*Nycticebus coucang)* changed from Near Threatened to Vulnerable because of pet trade harvesting and habitat loss, while the Choco Broad-nosed Bat (*Platyrrhinus chocoensis)* went from Vulnerable to Endangered. On the contrary, community conservation programs that protect specific roost sites helped elevate the Pemba Flying Fox (*Pteropusvo eltzkowi)* from Critically Endangered to Vulnerable (Altieri, 2013).

To determine the pattern in the global status of mammals and pollinating birds, the Red List Index (RLI) - a strategy that shows patterns in survival odds over the long term for species sets using data from The (IUCN) Red List. The ratio of species that move along the IUCN Red List categories across time, either away from or toward extinction, serves as the basis for the RLI. The method currently should be extended to include taxonomic groups contributing more significantly compared to vertebrates to pollination, like butterflies *(Lepidoptera)* and bees and wasps *(Hymenoptera)* (Lawrence, 2015).

Invasive Species

Figure 3.3: Some examples of flower pollinators: (**A**) hummingbird, (**B**) honeybee, and (**C**) butterfly pollination.

Source: Alyasseri ZAA, Al-Timemy AH, Abasi AK, Lavric A, Mohammed HJ, Takahashi H, Milhomens Filho JA, Campos M, Hazarbassanov RM, Yousefi S. A Hybrid Artificial Intelligence Model for Detecting Keratoconus. *Applied Sciences*. 2022; 12(24):12979. https://doi.org/10.3390/app122412979 [accessed 12 Jun, 2023]

Humans have been moving plants and animals from one section of the globe to another for 1000s of years. Sometimes they do it intentionally, like when they move livestock and companion animals like cats and dogs, but sometimes they do it by accident, like when rats escape from boats. Introduced species frequently fail to adapt to new environments in the wild. Nevertheless, a portion grows and spreads. These "invasive alien species" may have devastating effects on the fauna and flora of the area. Invasive species can consume native ones, compete with, hybridize, disrupt, or destroy their habitat, or, for pathogens or parasites, infect or kill native species. Over the past 500 years, invasive alien species have been partially or wholly responsible for the eradication of about 65 bird species, making this the most well-known contributory factor in the ongoing losses to the globe's avifauna. Intrusive species are a key issue for bird populations across the planet. These include plants, animals, or disorders that are non-native to a region and can destroy the local environment. Invasive species can spread disease, compete for food, and destroy habitats for birds. They may even directly target birds in some instances. For instance, many ground-nesting birds have perished as a result of rats consuming their eggs and chicks, which were introduced into island ecosystems. Climate change is aggravating the issue by expanding the scope of a few intrusive species. For instance, mosquitoes that transmit avian malaria can survive and breed at higher altitudes as Arctic temperatures rise, placing more birds at risk (Altieri, 2013).

Altogether, 156 bird species are acknowledged to become extinct since 1500. The most significant factors have been the effects of invasive alien species, habitat loss, and conversion, and overexploitation by humans caused by logging and agricultural spread. Roughly 109 species have become extinct as a result of invasive species. Predation by invasive rats and cats has become the most risky factor, contributing to the eradication of about 49 to 42 species, respectively. Predation by invasive canines, pigs, and mongooses, habitat loss by sheep, goats, and rabbits, and disorders brought about by introduced microorganisms have been implicated several times (Szabo et al. 2012).

The introduction of (*Boiga irregularis)* or brown tree snakes is a classic illustration of the devastating impact that invasive predators can have on ocean islands. The brown tree snake has destroyed bird populations on the Pacific island of Guam after it was coincidentally introduced after WWII (Rodda et al. 1998). 12 of the island's local bird species have perished, and 8 more have experienced significant population declines due to its exceptional predatory abilities and the absence of natural enemies on Guam (Wiles et al. 2003).

Recent observations of the snake on a few of the adjacent Northern Mariana Islands (Rodda and Savidage, 2007) have led to the reclassification of a number of species as Critically Endangered due to predictions of extremely rapid population drops in the near future (Camp et al., 2009). The Mariana Crow, *Corvus kubaryi*, is one species whose numbers have already shown a noticeable decline as a result of extensive habitat destruction (Plentovich et al. 2005).

3.4.2 Bats

The process by which bats pollinate plants is known as chiropterophily. Bat-pollinated plants typically have pale, nocturnal flowers, whereas bees prefer bright, daytime flowers. These flowers are much of the time enormous and bell-shaped, and a few bats have developed explicitly to reach the nectar at the lower part. In desert and tropical climates, bats are crucial pollinators. Southeast Asia, Africa, and the Pacific Islands are home to the majority of bats that visit flowers.

Figure 3.4: Bat pollination.

Source: Albuquerque-Lima S, Taylor NP, Zappi DC, Machado IC. Floral Specialization and Bat Pollination in Subtribe Cereinae (Cactaceae): A Morphological Approach. *Diversity*. 2023; 15(2):207. https://doi.org/10.3390/d15020207

However, bats are endangered as pollinators due to many factors, some of which include:

Loss of Habitat

The decline in bat numbers is a reflection of the rapidly-evolving countryside. Hedgerows, woodlands, and ponds are examples of natural habitats that are deteriorating and fragmenting. We should make new suitable habitats, and manage and improve existing environments to help bats recuperate and survive. The bats' reliance on insects as their sole source of food has diminished as a result of pesticide use, habitat loss, and intensive farming practices. For instance, the shift from making hay to silage has reduced the number of insects reaching adulthood. Climate change may also affect the life cycles of insects, which could affect bats' feeding trends (Altieri, 2013).

Building and Development Work

Many bat species perch in buildings and are rather vulnerable to human activities. Bats using buildings for habitat are currently threatened by building works if they are ongoing, while the work is in progress, or when destruction is occurring. In case bats are upset at an especially delicate season (for example during winter hibernation), it can negatively affect local bat populations.

Cat Attacks

One of the most common causes of bat fatalities is cat attacks. A cat's capture of a bat will almost certainly result in its injury. A bat caught by a cat will require assistance even if there are no visible injuries because there is a high risk of internal contamination if even a small volume of cat saliva infiltrates the bat's bloodstream. Despite not consuming bats, cats enjoy playing with them. Keeping your cat indoors at night can protect a whole colony because cats frequently learn where bats roost and will return back to catch bats as they leave (Lawrence, 2015).

Lighting

Bats are nocturnal animals that have learned to live in darkness, partially to avoid being eaten by birds of prey like sparrow hawks during the day. As a result, bats can be extremely disturbed by artificial lighting in their roosts, access points, and foraging routes. For bats, artificial lighting spread over or adjacent to bat roosts can cause various issues, including:

- Preventing or delaying roost emergence, which reduces foraging time and misses the insect abundance peak (right after dusk). Since all bats in the UK consume insects, this missed opportunity may have a significant impact on their capacity to thrive.
- At times leads to bats forsaking or becoming buried in the perch

when the entrance is illuminated since bats will be less disposed to leave their perches in these conditions. This is probably going to be viewed as a regulation breach since both bats and their perches are safeguarded by law.

- Affecting the feeding trends of bats far from the roost feed. Slow-flying species i.e. lesser horseshoes, 6 *Myotis* species and greater horseshoes keep away from enlightened regions and thusly lose feeding grounds when they are lit. The so-called "vacuum effect" occurs when slower-flying species are forced to use lower-quality foraging locations and lose out on prey that is drawn to the bright areas (Altieri, 2013).

White Nose Syndrome

Figure 3.5: White nose syndrome is a major threat to bat populations. Skin infection with white-nose syndrome fungus

Source: Marvin Moriarty/USFWS - This image originates from the National Digital Library of the United States Fish and Wildlife Service [accessed 12 Jun, 2023]

The condition known as white-nose syndrome is caused by the *Pseudogymnoascus destructans* fungi. The disorder is believed to have killed a huge number of bats in eastern North America beginning around 2006 and can eradicate up to 100 percent of bats within a colony. According to statistics provided by the US Fish & Wildlife Service, millions of bats have perished in 38 states and eight provinces across Canada due to the white-nose syndrome (WNS). Since the condition was first discovered in 2006, numbers have decreased by 80-100 percent in some hibernation areas (Lawrence, 2015).

White-nose condition is caused by the microorganism *Pseudogymnoascus destructans*. Since 2006, it is estimated that the disease has destroyed millions of bats across eastern North America. During hibernation, it can kill up to one hundred percent of a colony's bats. The little brown bat (*Myotisluci fugus*) was found to have the first known case of white-nose syndrome in King County, near North Bend, in March 2016. However, the condition has decimated bat numbers in eastern North America. In general, unlike bats in Eastern North America, those in Washington do not hibernate in large groups. Hence, the spread of the condition in Western states could be different.

During winter hibernation, the infected bat may develop a white, fuzzy appearance due to the fungus growing on its nose, wings, and ears. When the bats wake from hibernation, this fluffy white appearance may disappear. The fungus invades deep-skin tissues and causes severe damage, even though it may not be visible. Infected bats arouse more frequently during hibernation which makes them use critical fat supplies, prompting starvation and death. Damage to the wings, inability to control body temperature, difficulties breathing, and dehydration are additional disease-related causes of death (Altieri, 2013).

Case Study: The Mauritius Flying Fox

The controversial cull of the Mauritius flying fox, a fruit bat native to the island of Mauritius, has been a major cause of concern. These endangered bats have been killed in the thousands. Researchers are suing the Mauritius government to end the extermination of flying foxes. 50,000 bats have been killed thus far. The small island nation of Mauritius, which is east of Madagascar, is renowned for its picture-perfect beaches, hospitality, and cultural variety. Additionally, it is well-known for its dodo, the symbol of human-caused extinction. The Mauritian flying fox is currently on the verge of extinction. Officials assert that culling is required to safeguard fruit growers' crops, they have been accused of causing significant damage to plantations.

According to preliminary research, these flying foxes are responsible for 9% of the damage to lychees and 11% of the damage to large mango trees. Growers contribute to the waste by permitting their fruits to overripe, according to studies published in the Journal for Nature Conservation. Strong winds and other insects and animals that eat fruit compound the problem. The Food and Agricultural Research & Extension Institute of Mauritius claims that after the culls began, fruit yields have actually decreased. However, bats are still blamed and targeted.

Mauritius excels in conservation in some ways. In 1992, it became the first nation to approve the Convention on Biological Diversity. Since then, it has helped to save endangered species like the Mauritius kestrel, which now has roughly 500 individuals living within the zone. When wildlife is perceived to have an impact on economic interests, there is usually minimal political will, even for a government that is an international model for conservation (Lawrence, 2015).

Flying foxes spread seeds and pollinate flowers as they consume fruit, ensuring the survival of forests. The most significant illustration of a single bat species' contribution to a forest that scientists are aware of was discovered by Florens and his colleagues in a 2017 study, they found that fruit bats disperse seeds for 53% of Mauritius forest trees. These diverse trees are reliant on this one small endangered bat that the public is slaughtering. The bats' activities are particularly significant to Mauritius, where previous seed dispersers — such as dodos, broad-billed parrots, and giant turtles have gone extinct.

After fruit growers in Mauritius began exporting their lychees and mangos, which they had traditionally sold at local markets, in 2002, trouble started for the bats. Concerns about losses increased in tandem with profits. For the vast majority of fruit growers, flying foxes became suspect for chewing mangoes and lychees which sustain the fruit farming business. They thus became a target for eradication. It is estimated that 579,000 fruit bats were residing in Mauritius in 2015, however when the government began its first large-scale cull, officials quickly realized that hunting bats near farms was not a good way to meet the goal of reducing the number to 30,000. According to Florens, they would shoot one bat in the orchards, and more would flee. Consequently, they started looking for bats deep inside the forest, even in protected zones where the bats gather to roost. According to Florens, this strategy was "unusual" because it targeted bats in zones where they should be safest and removed some of the population's least troublesome bats—those likely not to consume farmers' fruit.

According to Mahen Kumar Seeruttun, a Mauritius member of the parliament, the government launched a second cull in 2016 of 7,380 bats " to further manage the damages done to fruit trees in a few selective orchards." The International Union for the Conservation of Nature (IUCN), which determines species' conservation status, decided in 2018 to upgrade the Mauritian flying fox's position from vulnerable to endangered as a result of this. The IUCN likewise coordinated two workshops facilitated by Mauritius' Ministry of Agro-Industry & Food Security to encourage nonlethal ways to lessen disputes between individuals and bats. However, shortly thereafter, the government initiated a third culling of 13,000 bats, which only recently came to an end (Altieri, 2013).

3.4.3 Butterflies

Butterflies are highly active during day time and visit different wildflowers. The flat, clustered flowers that provide a landing pad and abundant rewards are those that butterflies typically prefer to search for nectar, which provides energy for flight. Butterflies have great vision but a poor sense of smell. Butterflies, unlike bees, can see the color red.

Numerous butterflies emit scents that draw insects of the opposite gender. Large numbers of these fragrances frequently smell like the flowers that they are drawn to and visit. The scent of these flowers pollinated by butterflies may have developed as an adaptation that capitalizes on the scents' inherent attractiveness. The decline of butterfly populations is a constant threat. Among the most significant factors in species decline is habitat loss. The key dangers to butterflies are territory change and habitat losses because of private, business and agrarian development. Numerous species of butterflies are also in danger as a result of invasive species, climate change, and the widespread use of pesticides. These factors have both direct and indirect impacts on native host plants.

- **The Karner Blue Butterfly *(Plebejus samuelis)***: It is one of the world's most endangered butterflies coming from the U.S.A. In 1992, the federal government listed the Karner Blue butterfly as an endangered species. The wild lupine plant leaves are the only food for Karner Blue caterpillars. This seriously limits where they can thrive. The Karner blue butterfly's range has been lost due to land development and natural disturbances from wildfire and large mammal grazing. By pushing back on activities that are encroaching on the butterfly's habitat, and encouraging the growth of lupine and flowering plants, this can help to preserve the butterfly populations.

- **Mitchell's Satyr Butterfly (*Neonympha mitchellii mitchellii*)**: It is also found in the USA and was enlisted as endangered in 1992. The Mitchell's Satyr lives in low-nutrient wetland fens that receive carbonate-rich underground water from critically endangered seeps and springs. The majority of these wetland habitats have been drained and filled in to allow for agricultural and urban development. Likewise, attacks from exotic weeds compromise the fens whereby the butterflies depend. Pesticides, fertilizer, and nutrient runoff from neighboring agriculture—including livestock production—pose a threat to the butterfly population in fen wetlands. It is believed that butterfly collectors eradicated some of Mitchell's satyr populations (Lawrence, 2015).
- **Macedonian Grayling (*Pseudochazara cingovskii*)**: In 2010, the Macedonian Grayling was placed on the critically endangered list in Europe, the Republic of Macedonia. The Macedonian Grayling has a rather small range and only resides in Macedonia. Quarrying poses the greatest threat to it. The impact of quarrying on the existing and new populations are being actively investigated by researchers. The long-term success of conservation will depend on raising public awareness and collaborating with quarry owners.
- **Wolkberg Zulu Butterfly (*Alaena margaritacea*)**: Found in South Africa, The Wolkberg Zulu butterfly is an elusive species to inspect because it is so rare. The encroachment of humans and the loss of habitat by invasive species pose a threat to its population. The lichens that larvae consume are extremely sensitive to ecological changes.

Reasons for Decline

The variables causing the decline of butterflies fall into three primary classes: Chemical pollution, habitat degradation, and climate change—the latter of which has both positive and negative impacts depending on the species and location (Altieri, 2013).

- **Habitat Degradation**: The destruction and loss of habitat have unquestionably been the most significant factor in butterfly decline. Since the 1950s in the UK, there has been a 97% decrease in flowering meadows, an 80% decline in calcareous grassland, a 50% loss of ancient native forestland, and a 40% loss of marshland. Other nations in Western Europe and, to a lesser extent, the rest of Europe have suffered comparable losses. The expansion of intensive agriculture has resulted in the reseeding and fertilization of pastures and other habitats including the tilling of grasslands for cash crops, which is a major contributor to these losses. During the 20th century, large-scale and extensive farming practices replaced traditional activities like increased grazing and hay cutting in most of Western Europe. This prompted a significant loss of natural herbs and nectar sources that different butterfly species depended on. Some species may benefit from abandonment in the short term, but only a few tree- and shrub-feeding ones will benefit as the vegetation transitions to woodland.

Some butterflies breed in the forest canopy, feeding on deciduous trees, while others live in open areas of the forest: like in glades, and clearings. The replanting of non-native coniferous trees with their dense shade and the absence of traditional management are also contributing factors in the decline of these open space species. Up to the 20th century, most woodlands in Western Europe were coppiced, creating a variety of seral stages where different butterflies could breed. Numerous species, including *Boloria euphrosyne* and *Boloria selene*, have seen their numbers plummet as only a small ratio of the forest is actively coppiced.

The fact that the remaining habitats are comparatively small and isolated is one consequence of extensive habitat loss and degradation. Either through regular stochastic processes or inbreeding depression, populations breeding in such zones are highly likely to die out. Many butterflies, particularly sedentary habitat specialists, are concerned about habitat fragmentation. They often occur as metapopulations ranging in several small habitats in fragmented landscapes. Across Western Europe, habitat fragmentation has been extreme and is a major problem (Lawrence, 2015).

- **Chemical Pollution**: Chemical pollution includes a wide range of substances that harm butterflies and their natural environments. The most common are pesticides, which are regularly showered on arable yields to lessen destruction by insects and different organisms. Regardless of the expected risks of pesticides, minimal research has been conducted on butterflies. The greatest recent concern arises from neonicotinoids, chemical substances that were introduced during the mid-90s and that have been involved in the decline of honey bees and insectivorous birds (Altieri, 2013).

The effects of neonicotinoids on butterflies have not been extensively studied. In the laboratory, it is known that they kill Monarch butterflies, whereas in the field, they have been found in massive quantities in the host plants. Harmful impacts have likewise been shown for butterfly predators and different insects. Evidence comes from a research done in UK showing a correlation between the decline of "larger countryside" butterflies and the utilization of neonicotinoid pesticides. Nevertheless, this research did not establish cause and effect, and additional research is required on both direct and indirect effects of butterfly behavior, longevity, and reproduction that may be impacted by mortality. All EU crops are now prohibited from containing neonicotinoids, but adverse effects may continue in the coming years.

Aerial nitrogen deposition is another pollutant that harms butterflies. Ammonia from intensive livestock rearing and vehicle emissions of nitrogen oxides are the primary sources. Nitrogen pollution has been mentioned in the decline of butterflies since it changes either microclimates, or the vegetation where they breed. Specialty butterfly habitats typically lack nutrients, allowing for a wide range of plant species, including those that cater to butterflies and those that provide structural support. Nitrogen enrichment promotes the spread of nitrogen-tolerant species to the detriment of those that thrive in nutrient-poor conditions. Nitrogen deposition is believed to be responsible for the decimation of *Lasiommata megera* in the Netherlands since it supports vegetation growth and decreases exposed grounds where the butterflies breed. In the European Grassland Butterfly Indicator, this butterfly is among the species that is disappearing at the fastest rate.

A study demonstrates that butterfly species that winter as adults or chrysalids are declining at a faster rate compared to those that winter as eggs or caterpillars. This provides additional evidence that nitrogen deposition can cool otherwise warm microclimates. This is probably because caterpillars are becoming prey to the increasingly rapid growth of vegetation during the early spring, which lowers vegetation temperatures and, as a result, their rates of growth and survival chances (Lawrence, 2015).

- **Climate Change**: This can affect butterflies in both positive and negative ways. The scope of a few thermophilous species has spread rapidly toward the north in Europe as an immediate outcome of climate change, occasionally by a few hundred kilometers. Then again, cold-adapted species have withdrawn to cooler conditions higher up in the mountains or have gone locally extinct. Certain species that reside on mountain ridges are particularly undermined, considering a few endemic species found in the Sierra Nevada, South Spain.

  In Europe, the growth season has expanded by over 10 days starting from 1992, prompting faster development of natural vegetation and issues for species related to intermediate successional vegetation by impacting either their microclimate or the nutrient quality of food plants. *Voltinism* is additionally a significant, yet complex, factor. *Multivoltine* species are decreasing less quickly than univoltine ones in the UK, though the inverse is true for the Mediterranean area of Spain. One more issue brought about by climatic warming is that species with complex life cycles, for example, butterflies may experience the negative effects of developmental snares if ecological signs to enter diapause are upset (e.g., *L. megera* in Belgium) (Altieri, 2013).

  Climate change is anticipated to heighten the frequency of destructive weather events like floods and droughts. As mentioned before, the 1976 drought was followed by a quick drop in butterfly numbers in the UK, from which a few breeds have never recuperated. Correspondingly delayed rainfall can lessen breeding success and diminish population sizes. In fragmented regions, this may increase local extinction levels and influence the feasibility of meta-populations.

Conservation Solutions

Landscape-Scale Preservation

In Western Europe, numerous key habitats exist only as isolated and small fragments. For instance, seminatural grasslands presently contain under 0.6% of the land mass of Britain and Wales, and most fragments are under 10 ha in measurement. Most threatened species, presently exist on little areas of territory encompassed by intensively utilized land. To handle the issue, Butterfly Conservation has fostered an extensive landscape preservation exercise in the UK, focused on 200 priority landscapes for endangered butterflies and moths. These have demonstrated that declines can be stopped by improving land management site networks (Altieri, 2013).

Figure 3.6: Butterflies provide pollination services to farmers

Source: Saul-Gershenz L, Grodsky SM, Hernandez RR. Ecology of the Western Queen Butterfly *Danaus gilippus thersippus* (Lepidoptera: Nymphalidae) in the Mojave and Sonoran Deserts. *Insects*. 2020; 11(5):315. https://doi.org/10.3390/insects11050315

Landscape areas are selected to include site networks supporting meta-populations of endangered species. Around half of them are considered high priority since they support a) more threatened species; b) significant ratios of distributions of such threatened species; c) occupied and unoccupied site networks and d) more habitat types where management intervention can be embraced. Landscape-scale conservation includes three primary components: i) expanding the area of appropriate breeding habitat; ii) improving habitat quality through targeted management, both to upgrade existing populations of endangered species and to make unoccupied destinations appropriate for (re-)colonization; and iii) improving connectivity both inside and between locations — in this way assisting butterflies with movement and expanding the rate of colonization (Lawrence, 2015).

Chapter

4 ANIMAL HAZARDS IN THE ENVIRONMENT

CONTENTS

4.1 INTRODUCTION

One of the fundamental goals of animal health care is to boost livestock production. Reducing the incidence of fatal illnesses is usually the first step in increasing the number of cattle produced in developing countries. Fulfilling these aims could probably create damage and deterioration to the surrounding ecosystem due to increasing animal populations, poor management, and excessive consumption of scarce natural resources and veterinary drugs. If increased livestock output caused by development is not adequately managed, it may result in overgrazing, soil fertility loss, soil erosion, and finally desertification. These are all possible outcomes. The bulk of ruminants raised in intensive settings are likely to be located in agricultural zones with limited output potential. Vaccination and parasite control initiatives (for tsetse flies and ticks), while beneficial to production and population growth, may contribute to land degradation if not coupled with suitable land-use planning. Vaccination and parasite control efforts (for tsetse flies and ticks) have been demonstrated to improve productivity and population growth (Brown, 2021).

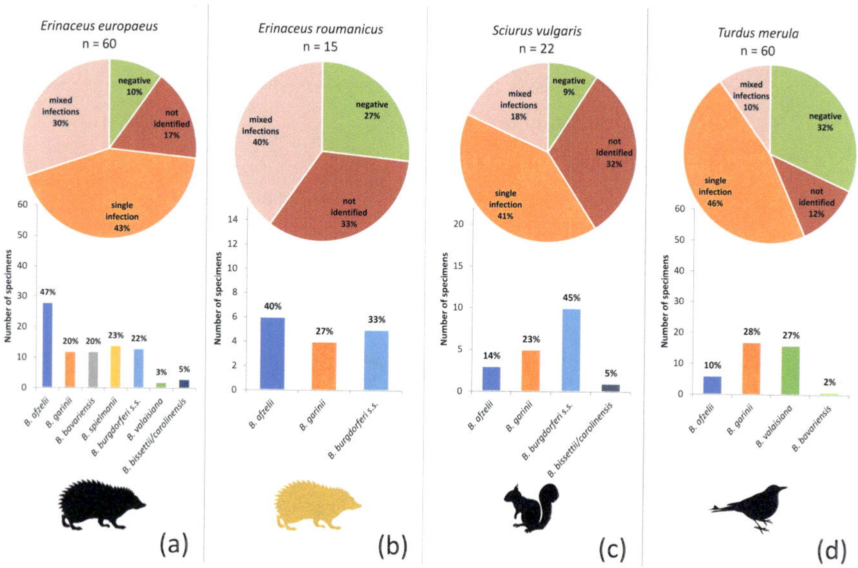

Figure 4.1: Sentinel animals as a warning system in the environment-animal-human web of interactions.

Source: Majerová K, Hönig V, Houda M, Papežík P, Fonville M, Sprong H, Rudenko N, Golovchenko M, Černá Bolfíková B, Hulva P, et al. Hedgehogs, Squirrels, and Blackbirds as Sentinel Hosts for Active Surveillance of *Bor-*

relia miyamotoi and *Borrelia burgdorferi* Complex in Urban and Rural Environments. *Microorganisms*. 2020; 8(12):1908. https://doi.org/10.3390/microorganisms8121908 [accessed 12 Jun, 2023]

It is critical to properly plan and exploit land use, taking into account the multiple agricultural, topographical, and geographical variables involved, to reduce the risk of adverse environmental changes, increase production, and reduce animal sickness. To attain these objectives, efficient land use planning and use are required. As a result, an interdisciplinary strategy is required to guarantee the most effective use of land and planning. A similar trend may be seen in the rising use of veterinary products such as insecticides, as well as the increased accumulation of residues such as feedlot dung as a result of animal production. Both of these phenomena are linked to an increase in the number of animals reared. When building new animal housing, especially in industrial production systems, it is critical to consider the growing problem of environmental pollution or contamination produced by animal wastes (manure and liquid manure). This is especially crucial to remember because the condition is predicted to worsen. It is critical to maintaining the requisite level of attention to guarantee that abattoir waste is properly used or disposed of securely. If these waste products are treated appropriately, they may have monetary worth. Any item that has been ruled unsafe for ingestion or has been proven to be contaminated must first be sterilized or converted before it may be reused. Inadequate waste disposal may increase the number of predatory species (such as hyenas, farm canines, and other animals of this type on land, and sharks in the water). Animal illnesses can be transmitted to humans through the ingestion of infected animal products; thus, all leftovers from foreign flights and ocean voyages must be sterilized (Brown, 2021).

It is now feasible to deliver acaricides (such as pour on) and insecticides (such as tsetse control targets and traps) in environmentally friendly methods. They should be used more regularly since they lessen the possibility of environmental contamination. The use of trypanotolerant cattle or tick-resistant breeds, as well as the adoption of other hybrids or strains that naturally resist parasites, can all help to reduce pesticide use (Wiedmann, 2011).

4.2 NATURE OF ANIMAL HAZARDS

4.2.1 Wildlife Attack

Animal attacks, which are violent crimes perpetrated by nonhuman animals against humans, usually entail biting the victim. Biting is the most common type of assault in animal attacks. As a result of these assaults, there have been casualties and deaths in a number of nations throughout the world. According to the United States Ownership and Demographics Sourcebook for 2012, 56. Over 4.7 million individuals in the United States were victims of dog attacks in 1994. Both geographical location and hormonal activity can influence the frequency with which animals attack. Both androgen and estrogen are hormones generated by the gonad glands, which are located in front of the pituitary gland. Animals with greater levels of these hormones are more prone to attack humans and other animals; depending on the species, this may occur more frequently at specific periods of the year, such as the rutting season. According to a National Safety Council study, a person in the United States is more likely to be killed by lightning than by their dog while they are at home.

Animal assaults are widely recognized as a hazard to public safety, and this issue has been considered. According to 1997 statistics, animal assaults on people occur at a rate of up to 2 million individuals each year in the United States. Every year, animals cause the deaths of hundreds of people all around the world. When a predatory animal approaches and assaults its primary target, such as when hunting humans, this is referred to as an unprovoked assault. Every year, the Centers for Disease Control and Prevention receive reports on all types of mortality. In the field of medicine, injury codes are used to designate specific conditions. The World Health Organization uses the same classification method; however, it is uncertain if all governments keep track of animal-related mortality. Even though, except a few tigers, animals only occasionally actively pursue humans, events like this one are alarming since they "harm the public image of many species."

A conflict between two animals is the most prevalent cause of animal bites. Every year in the United States, an estimated 250,000 people are bitten by dogs, 1-2 million people are bitten by cats, 45,000 people are bitten by snakes, etc.. Animals such as skunks, horses, squirrels, rats, rabbits, pigs, and monkeys can cause up to 1% of bite wounds. Pet ferrets have been known to attack their owners without warning, causing severe face injuries. The widely held assumption that wild animal bites may spread rabies is erroneous. Bites from wild animals account for fewer than 1% of all reported bites. When someone gets assaulted on their right arm, it is usually because they are fighting with their dominant arm, which is their stronger arm. The great majority of deadly bites occur on the head. Seventy-five percent of all human attacks are estimated to target the limbs or legs. Victims receive injury to their faces in only 10% of all attacks. The majority of bite victims are under the age of 10, and the majority of these victims are youngsters. Bite injuries can occur as a result of both human assaults and animal attacks. Other human assaults are the third most prevalent form, behind dog and cat attacks. Children are the most common victims of dog bites, which most often occur on the face and scalp (Brown, 2021).

Because of the probability of developing rabies and the presence of a variety of bacteria in animal buccal canals, the risk of infection following an animal bite is considerably enhanced. A handful of these disorders have been identified as a result of microbe research. These diseases are virtually often caused by a symbiotic relationship between aerobic and anaerobic microbes. *Pasturella* spp., Streptococcus, Staphylococcus, Moraxella, Corynebacterium, Neisseria, Fusobacterium, Bacteroides, Posphuomonas, *Capnocytophaga canimorsus*, and Prevotella are some of the bacteria that may be found by bite remains and exposure to other variables and shifts in the physical environment. Prevotella is another germ that can be found (Wiedmann, 2011).

4.2.2 Venomous Animals

Poisonous substances are produced by a wide range of animal species, which they may then employ to either kill or immobilize their prey to prevent it from being eaten by other animals. Envenoming is the process by which a venomous animal injects its poisons (also known as venom) into its prey via its venom mechanism, which might be teeth or a stinger. Dangerous animals, on the other hand, have the potential to passively poison their prey whenever they come into contact with them, whether by breathing, touching, or eating. Poisonous or venomous animals differ solely in the manner in which they deliver their toxin. Only the organisms that have been named are toxic. Venoms have developed for a variety of causes over time. Their intended effects might range from simply being unpleasant to being severe enough to cause paralysis or even death. They may opt to retain a small number of coevolved organisms while ignoring the variety of the rest of the biota. Something that kills one species might not have any effect on another. This list only includes animals whose venom has been proven to be harmful to humans or domestic animals. This is because the term "venomous" may have different implications depending on who you question.

Venomous organisms are classified into various phyla, each with its particular taxonomy, poisons, targets, clinical effects, and consequences.

Venomous snakes, the most dangerous type of snake, are responsible for up to 100,000 deaths each year, one and a quarter million injuries, and the lasting disability of many more individuals. Spiders, ticks, jellyfish, marine snails, octopuses, and fish can all cause envenoming; however, scorpion bites are the second most prevalent cause of envenoming. Ticks, fish, and jellyfish are some species that can envenom you.

Neurotoxicity that produces paralysis, neurotoxicity that activates the nervous system, myotoxicity, interference with coagulation, hemorrhagic activity, renal toxicity, and cardiac toxicity are some of the specific consequences. Headache, vomiting, stomach discomfort, hypertension, hypotension, cardiac arrhythmias and arrest, convulsions, collapse, and shock are all common side effects.

Even if there aren't many of them, there are a few first aid methods that are sure to be useful regardless of the type of envenoming or the organism that caused it. Even though pressure immobilization bandaging is indicated for non-necrotic or hemorrhagic species wounds, the injured limb is frequently immobilized first, followed by the patient as a whole. Soaking the injured region in very hot water is the best therapy for really severe saltwater burns. Antivenom is the most significant tool for the second kind of medical treatment, which includes numerous therapeutic techniques (both broad and focused). However, not all potentially lethal species have access to antivenom; the substance is scarce in some of the world's poorest locations; and, in many cases, the use of antivenom is guided by tradition rather than rigorous scientific study (Brown, 2021).

4.2.3 Allergies

A considerable number of people are unable to be in the same room as animals due to asthma or other allergic reactions. An allergic reaction to any animal is possible; however, animals with hair or feathers are more likely to produce such a reaction. The vast majority of persons who are allergic to animals, on the other hand, acquire their symptoms after having close contact with cats and dogs.

According to the Asthma and Allergy Foundation of America, up to 30% of allergy sufferers in the United States may also be allergic to cats and dogs (Wiedmann, 2011).

Cat allergies affect about twice as many people as dog allergies. Humans can be allergic to proteins that are entirely harmless to animals.

The Urine, the Feces, and the Dander

Dander is a kind of allergen found in animal fur, hair, and feathers. It is made up of very minute pieces of dead skin cells and has the potential to cause allergic reactions in some people. Dander is an animal waste product produced by a wide range of animals, including rats, guinea pigs, hamsters, rabbits, birds, cats, and dogs.

Allergens are the substances that cause allergic reactions. Exposure to pet hair isn't the only thing that might cause allergic responses. Saliva, salivary proteins, and urine can cause it, and it is also possible for allergens such as pollen and dust to spread.

Animals without fur, such as reptiles, frogs, and fish, are less likely to elicit an allergic reaction. This is due to the fact that they do not shed their skin. The number of people who are allergic to pets, as well as the number of people who own pets, is increasing. People are getting pets in greater numbers.

According to the American Academy of Allergy, Asthma, and Immunology, there are more than 161 million cats and dogs in the United States who share their homes with their human families. When an allergen induces nasal inflammation, a disease called as allergic rhinitis develops. Dust, pollen, and mold are examples of common allergies. In allergic rhinitis, the body's immune system misidentifies a harmless allergen, such as a specific animal protein, as a dangerous foreign invader. This causes nasal and throat irritation.

People who are allergic to animal proteins may develop immunological reactions if they inhale or come into direct touch with the allergens (Wiedmann, 2011). When the immune system comes into contact with allergens, it reacts by releasing histamine into the bloodstream. Histamine is a naturally occurring chemical in the body that is responsible for the symptoms associated with allergic reactions. Benadryl, Zyrtec, and other antihistamine-containing drugs may be able to offer allergy patients some relief from their symptoms (Brown, 2021).

People who are allergic to animals are more likely to develop allergic contact dermatitis, a skin condition. This type of dermatitis is caused by an allergic reaction that occurs when an allergen comes into touch with an affected person's skin. Animal proteins can be carried in the air by small particles. Danger can remain in the air for long periods of time and stick to people's clothing as well as furniture. Animal saliva that has dried can stick to surfaces and then be released into the air. This can happen on rugs, clothing, and many other items. The amount of fur shed by ferrets and weasels is comparable to that shed by cats and dogs. This means they are hypoallergenic, which is great news for those of us who have pet allergies. If nurtured properly, ferrets have the potential to become pleasant, friendly pets that can even be housetrained. If you are unsure how to properly care for a ferret, see a certified veterinarian.

Various Types of Amphibians and Reptiles

Reptiles and amphibians are excellent choices for pet owners since they are hypoallergenic and visually appealing. Pets from the reptile and amphibian families include an iguana, a snake, a turtle, a leopard gecko, a salamander, a small lizard, a bearded dragon, and a frog. Some of these Animals , such as and mice, are likely to require certain lighting and heating arrangements to survive. Do you have a favorite amphibian or reptile you'd want to learn more about? If you require any additional information, please contact your local animal hospital.

Arachnids

Keeping a spider, tarantula, or scorpion as a pet isn't for everyone, but with the proper care and attention, any of these critters can become an intriguing and appealing addition to the family. They are also entirely hypoallergenic, which means they will not cause any allergic reactions if they come into contact with them. If you are thinking about keeping a spider or insect as a pet, you should get the advice of your local veterinarian first (Wiedmann, 2011).

Fish

Although fish cannot be caressed or carried around the neighborhood like other animals, they make fascinating pets since they are so engaging to observe and care for. Because the water in the aquarium keeps them submerged, they do not pose a problem to anyone with allergies. The process of building and decorating your fish tank could be a lot of fun.

Crabs on Their Initiative

The hermit crab is a simple pet that should not cause adverse reactions in the majority of its owners. Because they are usually kept in a tank or terrarium, the substrate located at the bottom of the aquarium should not be a source of worry. Hermit crabs are commonly referred to as "starter pets" for young children because they do not require much care and have unique patterns and colors on their shells.

4.2.4 Marine Animals

A variety of causes contribute to the dangerous condition of the global coastal waterways. Tides, currents, and rip currents all enhance the probability of drowning. Animals and plants both can damage people if they want to attack us, defend themselves against us, or come into contact with us by accident. If an animal bites, pierces, or stings you, it may use its teeth, spines, or stinging cells to inject venom or poison into your body. The most common marine animals that cause injury are jellyfish, fire coral, stonefish, sea anemones, sea urchins, and stony coral. Getting attacked by a shark, drowning, and getting stung by a box of jellyfish are the three most likely causes of death. There is no evidence that tourists are endangered by blue-ringed octopuses, cone snails, or sea snakes. Both huge and goliath groupers prefer the warmer waters of the tropics, where they can grow to be 3 meters (10 feet) long and weigh more than 300 kilograms (660 pounds). Divers risk being devoured by groupers as well as being paralyzed by head-on collisions, which are especially probable near sunken ships (Brown,. 2021).

Potentially Deadly Sponge Transporters

Seafloor organisms can be found anywhere in the ocean, from the shallowest seas and rocky shores to the deepest portions. Spicules of some species contain skin-irritating chemicals. The sea grass and oyster beds, as well as the rocks and trash where touch-me-not and fire sponges live, are popular tourist destinations in the Bahamas, Caribbean, the east coast of the United States, and Nova Scotia, Canada. Touch-me-not and fire sponges can also be found in the sea grass and oyster beds. There are several predatory marine animals, including man-of-war, jellyfish, fire coral, and hydroids. Nematocysts are stinging cells found in many marine animals, including jellyfish, man-of-war, fire coral, hydroids, and scorpionfish. When the nematocyst comes into contact with its prey or with humans, it releases a hollow, barbed dart that injects poison into the epidermis of its victim. Despite this, certain species' envenomation has the potential to be lethal. Many species' venom is either ineffective or the length of their darts is insufficient to do obvious harm to humans (Wiedmann, 2011). However, hydroids are not limited to garbage; they can also be found in caves, on reefs, and in plankton-rich coastal waters. They resemble grasses with tufts, fronds, feathers, fans, or even algae growing on them. There are a variety of plants and animals that can be dangerous to people in the Mediterranean and other warm-water regions. Blistering, redness, swelling, soreness, and a burning feeling in the affected area are all indications of envenomation.

Fire corals, like stony corals, have a brittle structure and a muted coloring that is frequently a drab mustard yellow, or light brown. Tropical and subtropical waters are home to fire corals. Because of the ubiquity of these organisms and their appearance as perfectly innocuous just below the surface of shallow water, many swimmers, unskilled snorkelers, and divers hopping into and out of boats on shallow reef edges wind up seriously hurt. Toxic species can be found in any region with tropical reefs that is not Hawaii. In addition to the intense agony, envenomation causes burning, redness, and swelling that can last for weeks.

The swimming bells are a little relative of the man-of-war that can levitate but cannot swim. Swimming chimes. The Portuguese man-of-war travels with the current and wind, thanks to its trailing, invisible tentacles covered in nematocysts and its visible, gas-filled, bluish-purple float measuring 30 centimeters (one foot) in diameter, but not in the Mediterranean Sea. This is due to the Portuguese man-of-war's unseen tentacles covered in nematocysts. Large wind colonies can grow and drift towards the coast; the appearance of these wind colonies near the coast may be recurrent. The Indo-Pacific region is home to a diverse range of animal life, including gentler species like the blue bottle. Envenomation causes the skin to produce localized wheals and blisters, severe pain, and eventually death. Feeling nauseous, spitting up, chest discomfort, difficulty breathing, and passing out are some of the warning signs that someone is likely to drown. There have been fatalities on many Caribbean islands and all along the Atlantic coast of the United States (particularly in the state of Florida) (Wiedmann, 2011).

There are around 250 different species of jellyfish, which can be differentiated by their color, size, form, and tentacle arrangement. All jellyfish are referred to as nematocysts. Jellies live in pockets on the ocean floor in every ocean. Genuine bell jellyfish tentacles are uniformly scattered throughout the jellyfish's body. Even among the same species, there is a wide range in envenomation severity. Numerous insect species can produce stings that are as painful as those delivered by the Portuguese man-of-war, but only very rarely result in fatalities. Four tentacles protrude from each of the four corners of a sea wasp or box jellyfish. They can only swim for around 5 feet, or about 1.5 meters, at a time. Two of the most dangerous jellyfish species include the Australian box jellyfish (*Chironex fleckeri*) and the Irukandji jellyfish (*Carukia barnesi*). The former can reach a maximum length of 2 meters (6.6 feet) and has a diameter of only 1.3 centimeters (0.5 inches), whilst the latter can reach a maximum length of 1.3 centimeters (0.5 inches). The Irukandji Jellyfish is rather widespread, with a breeding season that lasts from October to May (though some individuals are present all year) and spawns on reefs. In contrast, Australian box jellyfish are only found in tropical ecosystems and can be found along Australia's northern coast, which spans from Korea to India. The comparatively innocuous box jellyfish can be found in both tropical and temperate oceans across the world.

Irukandji envenomation symptoms include a broad rash, fever, and acute muscle pain; however, death from the illness is extremely rare. If first aid is not administered, envenomation from an Australian box jellyfish causes searing and horrible agony, wheals, blisters, muscle stiffness, and coma within minutes, and often results in death if the sufferer does not receive medical attention (Brown, 2021).

Coral

Coral is a complex creature composed of numerous little invertebrate animals, each of which deposits its calcareous skeleton. Coral reefs are home to a wide range of potentially dangerous organisms, such as animals, plants, and bacteria. True corals rarely sting, but fire coral can cause damage to the skin's surface layer. This coral has a very sharp edge, so avoid coming into contact with it. Coral wounds are famously slow to heal, may contain coral spicules or soft tissues such as nematocysts, and can result in bacterial infections. Coral wounds are not only prevalent but also famously difficult to heal. When delicate spicules found in soft coral skeletons come into touch with human skin for an extended period, they have the potential to penetrate the epidermis and cause an unpleasant rash.

4.2.5 Zoonotic Diseases

Zoonoses or zoonotic diseases are diseases caused by these infectious organisms. Bacteria, fungi, parasites, or viruses may be responsible for zoonotic disease transmission from animals to humans. These viruses can cause symptoms ranging from mild tiredness to catastrophic illnesses such as organ failure and even death in humans and other animals. Animals that carry infections that can infect humans may not display any symptoms of illness, however, this is very dependent on the individual zoonotic disease. Zoonotic diseases are a major concern in the United States and around the world. It is thought that three out of every four newly emerging or emerging infectious diseases, as well as six out of every ten known infectious diseases, can be transmitted from animals to humans. As a result, the Centers for Disease Control and Prevention (CDC) works around the clock to protect humans in the United States and around the world against zoonotic diseases (Wiedmann, 2011).

Because human and animal populations are so intertwined, it is critical to have a firm understanding of the common transmission mechanisms that zoonotic diseases employ to spread to people. The following are a few examples:

When a human comes into touch with the secretions, blood, urine, mucus, feces, or other body fluids of a diseased animal, that person is said to have had direct contact with an infected animal. Interactions between animals, interactions between animals and humans, and interactions between humans and animals are examples of this.

Coming into contact with contaminated objects or surfaces, as well as being close to an area where animals are present and active, are examples of indirect contact. Look for water in places like aquariums, pet bowls, poultry coops, barns, plants, mud, and dishes used for feeding and watering pets.

The term "vector-borne" refers to diseases spread by insects like ticks, fleas, and mosquitoes.

Every year, one out of every six persons in the United States becomes ill as a result of contaminated food consuming potentially hazardous goods such as raw milk (unpasteurized milk), raw meat, eggs, or fruit, as well as items contaminated with animal excrement. Polluted food can make humans and animals, even domesticated pets, sick (Brown, 2021). You will become ill if you drink water polluted with animal feces or swim in water contaminated with animal feces.

Meat, dairy products, and livestock byproducts such as flavorings, oils, and stocks have all been linked to the spread of illness. According to the Lancet, practices such as confinement, species mixing, feeding and slaughtering practices, and ineffective disease management all contribute to the spread of zoonotic diseases among livestock and their potential role as a source of novel infections in human populations. The procedures used for packaging, transporting, and keeping commodities may play a role in the spread of infectious diseases. Although annual meat consumption in Central

African countries is predicted to surpass 1 billion kilograms, consumption of wild commodities for subsistence purposes is still far lower than that of domestic cattle globally. According to The Lancet, increased cow production in countries with insufficient disease management practices may result in the introduction of new diseases, making it more difficult to reduce demand. This would make reducing demand more difficult (Wiedmann, M. 2011).

4.2.6 Agricultural Chemicals Used in Livestock Production

The most serious health concerns linked with a profession in agriculture are zoonotic illnesses, pesticide poisoning, work accidents, and respiratory disorders caused by organic and vegetable dust. Farmers are more likely to acquire skin cancer and dermatitis as a result of their constant exposure to the sun. They are placed in hazardous conditions where they are exposed to noxious fumes, animal allergies, and clouds of soil, and dust. Farmers who fail to monitor the accumulation of nitrogen oxide in their silos are more likely to die from silo-filler's illness, a farmer's lung is an allergy to moldy hay, a milker's knee is arthritis caused by years of crouching, etc. (Brown, 2021).

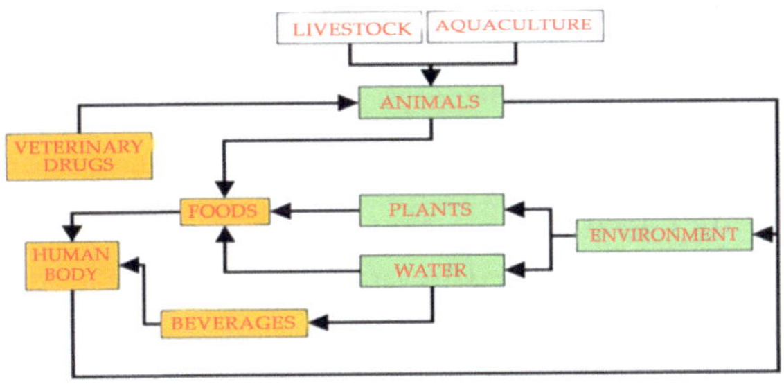

Figure 4.2: Veterinary drug and chemical residues in food of animal origin.

Source: Wang B, Xie K, Lee K. Veterinary Drug Residues in Animal-Derived Foods: Sample Preparation and Analytical Methods. *Foods*. 2021; 10(3):555. https://doi.org/10.3390/foods10030555

Those who run agricultural machinery represent the greatest risk to the general population health status on large-scale state and cooperative farms. A considerable proportion of persons suffer deafness as a direct effect of mechanical noise. Rollover accidents involving tractors are the leading cause of death on farms. Accidents in agricultural settings are not a new phenomenon, but as agricultural technology advances, the problem is becoming more prevalent in some countries. However, due to the low level of mechanization in a small-farm economy, each worker is expected to do a wide range of tasks. As a result, the task is both physically demanding and devoid of any feeling of specialization.

The breadth of product diversity and the amount of technological innovation are both indicators of the agriculture sector's vitality. Agriculture is distinguished by its cyclical structure and the requirement of various forms of field labor depending on the crop being farmed. This is true independent of the farm's size or the intricacy of the farming business. As a result, one of the main principles of occupational physiology, the normal workday, is routinely overlooked. Working long or irregular hours can result in overexertion, poor productivity, and fatigue, all of which can compromise an individual's natural resistance to illness. As a result, the individual may be more vulnerable to sickness. Pet animals are typically the offenders responsible for introducing undesired diseases and vermin into a home. Disease transmission has been observed in several countries, including brucellosis, TB, and others. Furthermore, tularemia instances have been observed in a wide range of geographical areas. Tetanus and other agriculturally-related diseases are extremely common (Wiedmann, 2011).

Although, acute poisonings are extremely rare, pesticide use has been related to cases of chronic intoxication in many countries. Toxin exposure is commonly underreported because even minor toxin exposure can cause flu-like symptoms or stomach pain. Pesticide exposure is thought to be to blame for both the fall in male fertility and the rise in the prevalence of allergy disorders after 1940. Those who work in agriculture and agricultural processing are routinely exposed to organic dusts made up of vegetables and other plant materials. This exposure has been associated with many occupational ailments, including byssinosis, farmers' lung, bagassosis, and occupational asthma, which must be reported on official disease registries in some countries. Such reporting obligations do not exist in other nations. Numerous dusts may be hazardous to human health; however, this has not been properly explored. Pollen from wheat, rice, cocoa, coconut fibers, tea, kapok, tobacco, and wood in farming contexts has the potential to aggravate obstructive lung illness and asthma in vulnerable persons. The flax dust that the people of this village have been inhaling since they were children has created lung difficulties in some of them (Brown, 2021).

4.3 DISTRIBUTION OF ANIMAL HAZARDS

Distribution patterns, an animal's surroundings, and its level of human interaction are all factors in determining an animal's proclivity to cause harm. This article contains crucial information on the geographic area of probable animal hazards. Dangerous animals can be found throughout the world, however, some are more abundant in specific areas than others. North and South America are good habitats for rattlesnakes, while Southeast Asia is home to venomous snakes like cobras and other dangerous species. Knowledge of these species' native ranges is extremely beneficial for assessing potential threats in specific locations of the world.

Animals have evolved to be able to survive in a variety of habitats, many of which are hostile. If certain marine organisms come into contact with people living in oceans and coastal areas, they can be seriously harmed or killed. It is easy to avoid potentially risky areas if one is familiar with the habitats chosen by various animal species. The quantity of interaction that occurs between humans and animals determines the chance of experiencing animal risks. People are more likely to come into contact with dangerous animals in locations where human activities interfere with the natural ecosystems. Wild animals are more likely to be encountered in places that have experienced extensive agricultural practices, broad deforestation, and widespread urbanization.

Climate and other environmental factors influence where deadly animals are most likely to live. While some species can only live and thrive in a narrow temperature range, others have adapted to a wide range of various settings and can do so with reasonable ease. Climate and environmental changes have the potential to influence the spread of dangers by influencing the behavior and distribution of various animal species. Human activities such as travel and commerce may have an impact on the location of animal hazards. As a result of globalization, animals and the insects that transfer disease to them, known as vectors, are allowed to roam freely between countries, offering new hazards in previously undiscovered locales (Wiedmann, 2011).

Ecological interactions can cause one species to influence the spread of another. Predatory or competitive animals, for example, may inhibit the spread of invasive species in specific parts of the world. The abundance of prey species can also influence the geographic distribution of predators, particularly those that represent a threat to people. Throughout the year, certain hazardous species follow migration and seasonality patterns. During the mating season, it is advised to avoid coming too close to the nests of certain bird species since they can become highly hostile. People are more likely to come into touch with migratory wildlife such as sharks and crocodiles since these animals tend to congregate in certain areas along their routes (Brown, 2021).

Agriculture, dam construction, and land expansion are all examples of environmental changes that could influence the proliferation of animal threats. Both the natural ecosystems on which animals rely and the habitats in which they reside are vulnerable to the effects of human intervention. Human-introduced invasive species can be hazardous to ecosystems because they outcompete native species and disrupt entire food webs. Many species spread to new places, posing a threat to agriculture and human health. Some of the species that invade new locations, for example, can be hostile or poisonous. Cultural and economic considerations may play a role in determining where potentially deadly animals live. A person's cultural upbringing and community conventions may influence how they react when presented with a potentially deadly animal. Communities may be more tolerant of the fact that a species is highly valued or legally protected if it has the potential to be destructive.

4.3.1 Agriculture

Agriculture is among the most common ways in which animal hazards are distributed. In as much as it contributes to the food given to the population, if not properly managed it can lead to drastic spread of animal hazards. Agriculture has the potential to increase the prevalence of animal dangers in many ways. In today's agricultural system, large herds of cattle are sometimes confined to restricted locations, such as feedlots or intensive agricultural operations. The likelihood of disease transmission between animals rises when there are a big number of them in one area. A sick animal can quickly transmit illness to the remainder of the herd or flock, as well as other farms in the area (Wiedmann, 2011).

Transporting animals is critical in the agricultural business for a variety of purposes, including breeding, trading, and animal disposition. The movement of sick animals may contribute to the development of animal hazards, since these risks may spread to new areas while in transit. The risk of disease transmission is heightened by the likelihood that animals brought in from other locations may be afflicted with illnesses that are significantly less frequent in the region. Agriculture generates a wide range of byproducts, including manure and other animal waste. There is a danger that the ecosystem, including the land, water, and air, may become contaminated if effective manure management is not maintained. Viruses identified in animal feces have the potential to infect other species as well as people if the waste is allowed to pollute water sources, be dumped into the environment, or remain dormant in the soil (Brown, M2021).

Agricultural methods that bring domestic and wild animals together may increase the likelihood of disease transmission between the two groups of animals. Cattle that are permitted to graze on pastures and ranges put themselves at risk of contracting infectious illnesses from the wild animals with whom they come into contact. Because of this link, there is a possibility of viral transfer from cattle to wildlife, which might lead to disease epidemics. Crop production and distribution are heavily dependent on global supply chains. Live animals, animal products, and plant components may all spread animal-related risks over the world. This includes the possibility of plant components transmitting animal-related hazards. Furthermore, insects and other disease-carrying vectors can spread disease across international boundaries on agricultural goods.

Antibiotics may be used in agricultural operations on occasion to accelerate animal development or to prevent disease spread. However, if medicines are taken wrongly or in excess, antibiotic-resistant bacteria may emerge. These bacteria are dangerous not only to people but also to animals. Through the adoption of initiatives such as biosecurity measures, vaccination programs, disease monitoring, and other management tactics, the agricultural sector plays a significant role in reducing threats to animal populations. If these safeguards are not properly implemented in an agricultural environment, or if gaps in monitoring and administration exist, this might foster the spread of animal dangers that are dangerous to humans (Wiedmann, 2011).

Crop-animals is typical practice in agriculture to raise crops and animals together. This may have some good consequences, such as greater biodiversity and nutrient cycling, but it may also increase the likelihood of disease transmission. Cattle that were allowed to graze in fields or on agricultural waste may have picked up viruses from the environment or the crops themselves. Farmers must use water management measures such as irrigation, water storage, and drainage systems to maximize agricultural production. These water management strategies could, however, result in the establishment and spread of dangerous species if they are not carefully planned and implemented. The West Nile virus and malaria are two illnesses that mosquitoes may spread to animals. These illnesses may be spread by stagnant water or poor irrigation methods (Brown, 2021).

Pesticides are often employed to eradicate insects and illnesses that might cause crop damage if left untreated. Pesticides, on the other hand, may have unforeseen effects if administered incorrectly or in excess. Some pesticides may provide indirect risks to animals in agricultural systems because they have the potential to damage non-target species, disrupt ecosystems, or induce animal impairments. These risks may be induced by insecticides that harm non-target species. Farms that focus on a few species of high-yielding crops or animals that are genetically stable are becoming more uncommon in today's agricultural scene. Plants and animals may be more prone to disease due to genetic similarities. If a new disease emerges that is capable of penetrating the immune systems of these genetically connected plants or animals, it has the potential to produce a widespread pandemic.

Agriculture is a key contributor to a variety of environmental challenges, including deforestation, soil degradation, and the loss of many species habitats. These changes may have an impact on wild animal populations as well as the relationships between those species and farmed animals. Changes in the environment, as well as changes in animal behavior, may have an influence on the contagiousness of illnesses in a wide spectrum of organisms, including humans. Climate change is expected to have an impact on agriculture due to changes in precipitation patterns, temperature rises, and severe weather events. These changes may have an impact on the rate at which animal illnesses spread as well as the places where they disseminate. Increased temperatures, for example, may encourage disease-carrying vectors to travel farther, but increased humidity may hasten the development of some diseases.

Despite the danger of animal illnesses spreading, agriculture is essential for human survival, economic growth, and quality of life enhancement. It is feasible to reduce the impact of animal threats while ensuring the safety and longevity of agricultural systems by using suitable mitigation measures, the most effective agricultural techniques, and efficient disease control procedures. This is made feasible by the use of optimum farming techniques (Brown, 2021).

4.3.2 Tourism

Tourism may contribute to the spread of animal dangers in a variety of ways. Visitors often come into touch with wild animals on safaris, wildlife reserves, and marine life observation cruises. Although these possibilities are intriguing, there is a possibility that will have a negative impact on animal behavior and the natural ecosystems in which they live. When people and animals are in close contact with one another, the chance of disease transmission increases. Vacationers may unwittingly carry infections with them or transmit diseases to others via contact with animals or environmental toxins (Wiedmann, 2011).

As tourism grows, there may be a larger demand for wildlife-related objects and activities, which might contribute to a rise in wildlife trafficking. The trade-in wild animals for the sake of tourism has the potential to significantly impact disease spread. People who come into touch with captured wild animals are at risk of catching a virus. It is likely that the circumstances under which these animals are kept, transported, and displayed contribute to the spread of disease. Travelers may unintentionally introduce non-native animals into a region by bringing them in their baggage. This happens when people release exotic pets into the wild or when humans move animals for tourism. The introduction of invasive species has the potential to harm an ecosystem's general health as well as its biodiversity.

The negative consequences of tourism on the environment may be divided into two categories: pollution and habitat destruction. Plastics, chemicals, and garbage generated by tourism-related activities are examples of pollutant types that may damage species and alter ecosystems. Wildlife population migration or fragmentation caused by the removal of natural habitats for the construction of tourism infrastructure might raise the risk of disease transmission among animals. The relocation or fragmentation of animal populations caused by the loss of natural habitats for the creation of tourism infrastructure may cause stress, congestion, or disturbance of ecological interactions. Tourists often feed wild animals by dropping food or delivering it directly to the animals, which may be done purposefully or accidentally. It is feasible that animals that rely on people for food would adjust their behavior and create greater ties with humans as a consequence. When wild animals are fed, they may assemble in large numbers, increasing the danger of disease transmission to people (Brown, 2021).

Chapter

5 LOSS OF BIODIVERSITY: CONCERNS AND THREATS

CONTENTS

5.1 INTRODUCTION

The word "biodiversity," which is derived from the phrase "biological diversity," refers to the astounding diversity of life on Earth at all levels of an organization, from the tiniest bacteria to the grand ecosystems that cover enormous regions. It is a fundamental feature of our world and includes the diversity of species, habitats, and genetic variation within species.

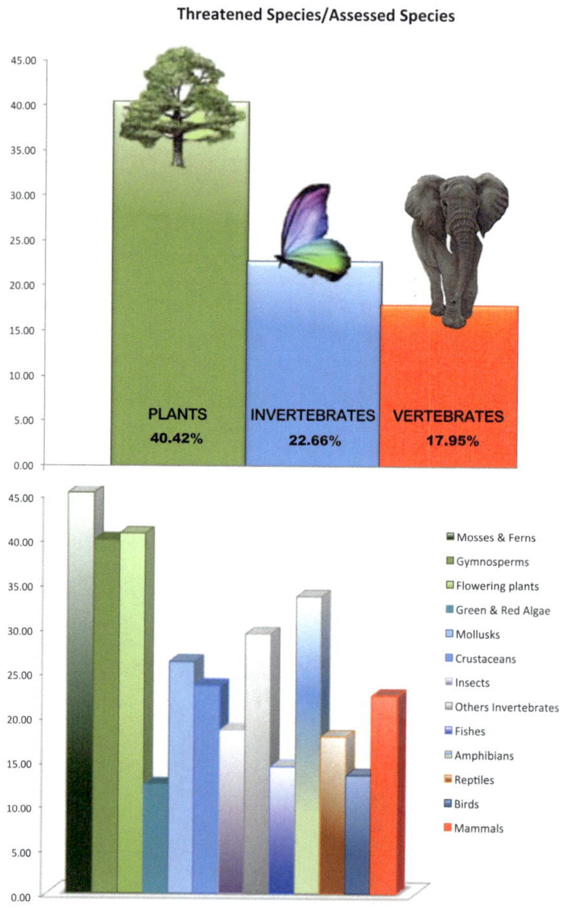

Figure 5.1: Percentage of the current number of threatened species evaluated in relation to the overall number of species assessed in the major groups of organisms by the IUCN animal Red list.

Source: Palombo MR. Thinking about the Biodiversity Loss in This Changing World. *Geosciences*. 2021; 11(9):370. https://doi.org/10.3390/geosciences11090370 [accessed 8 Jun, 2023]

All living things, including humans, depend on biodiversity, the intricate web of life that includes the astounding variety of species, ecosystems, and genetic diversity on our planet. However, we are currently facing a severe crisis as biodiversity is rapidly diminishing at an alarming rate. The loss of biodiversity has become a global concern, posing significant threats to the delicate balance of ecosystems and the overall health of our planet (Agarwal, 2015).

Biodiversity encompasses the richness and diversity of life at all levels, from the diversity of ecosystems, including forests, oceans, grasslands, and wetlands, to the immense variety of species that inhabit these habitats, and the genetic variation within each species. The interdependencies and connections found in this intricate web of life are essential to the health of ecosystems, the delivery of essential ecosystem services, and the sustainability of life on Earth.

Unfortunately, the loss of biodiversity is now primarily caused by human activity. Natural ecosystems have been severely disrupted by the rapid growth of human populations, industrialization, urbanization, unsustainable agricultural practices, deforestation, pollution, climate change, and the introduction of invasive species. These actions have caused habitat degradation, destruction, and fragmentation, which have contributed to the extinction of many species and the loss of genetic diversity (Kaltenborn, 2018).

The effects of biodiversity loss are widespread and have a significant impact on future generations, human societies, and the environment. The disruption of ecosystems weakens their resilience and impairs their capacity to deliver vital functions like pollination, water purification, nutrient cycling, and climate regulation. The loss of biodiversity also challenges food security, as many crops rely on pollinators, and the depletion of wild fisheries impacts the lives of coastal communities.

Furthermore, the loss of biodiversity threatens the possibility of scientific and medical discoveries, as many of our most potent medications and technological advancements have been taken from nature. Cultural diversity is also at stake, as indigenous peoples and local communities closely intertwined with their unique environments face the loss of traditional knowledge, practices, and spiritual connections to the natural world.

Examining case studies from diverse ecosystems around the world demonstrates the severe effects of biodiversity decrease, which helps us understand the scope of the loss. Coral reefs, tropical rainforests, freshwater habitats, and Polar Regions are all suffering from considerable biodiversity loss, which puts their complex ecosystems and the myriad species that depend on them in jeopardy (Agarwal, 2015).

Action must be taken quickly and cooperatively to address the loss of biodiversity. Protecting endangered species, recovering ecosystems, and advancing sustainable land and resource management techniques are all made possible by conservation and restoration activities. Conserving biodiversity requires global collaboration, policy changes, and active community participation.

Because it affects the health of our world and the well-being of all living things, the loss of biodiversity has received a lot of attention. As species disappear, environments deteriorate, and genetic variation declines, the intricate web of life that sustains us all becomes increasingly vulnerable. This loss of biodiversity poses a serious threat to the stability of our planet's ecosystems, the provision of essential ecosystem services, and the operation of ecosystems (Kaltenborn, 2018).

The disturbance of ecosystem services and processes is one of the main issues related to biodiversity loss. In complex and dynamic systems called ecosystems, species interact with one another and with their surroundings. Each species has a special responsibility for preserving the ecosystem's functionality and balance. A loss of ecological resilience and stability results from the delicate linkages and interdependencies within ecosystems that are disrupted as species disappear. Throughout the entire ecosystem, this disruption may have cascade impacts that affect everything from nutrient cycling and energy flow to predator-prey interactions and plant pollination.

The destruction of vital ecosystem services is a further cause for concern. Clean air and water, fertile soil for agriculture, climate regulation, flood control, and the provision of food and natural resources are just a few of the many services that ecosystems offer that are essential for human well-being. The loss of biodiversity threatens ecosystems' ability to provide essential services, endangering people's health, way of life, and economies. For instance, the decline of forests can affect the quantity and quality of water, while the extinction of pollinators like bees and butterflies can threaten crop output and food security.

Ethical issues are also raised by the destruction of biodiversity. Every species has a natural value and a right to exist on Earth. Future generations will be unable to appreciate and profit from the diversity of life as a result of the rapid loss of species brought on by human activity. Many indigenous tribes and cultures also have strong spiritual, cultural, and traditional ties to certain species and habitats. These crucial linkages between communities and their natural surroundings are disrupted by the loss of biodiversity, which also poses a threat to the preservation of cultural heritage (Agarwal, 2015).

The worries and dangers associated with biodiversity loss are made worse by climate change. Extreme weather events, altering precipitation patterns, and rising temperatures add to the stress already placed on ecosystems and species. The inability of several species to migrate or adapt to new habitats may result in higher extinction rates. Additionally, the decline in biodiversity jeopardizes ecosystems' capacity to act as carbon sinks and control the climate, aggravating the effects of climate change and setting up a risky feedback cycle. The decline in biodiversity affects both the environment and the economy. Agriculture, fishing, forestry, and tourism are just a few of the many industries that strongly rely on biodiversity and healthy ecosystems. Key species extinctions and ecological deterioration can cause financial losses, job insecurity, and fewer prospects for sustainable development.

At the local, national, and international levels, immediate and coordinated action is required to address the issues and challenges brought on by the loss of biodiversity. Stopping the loss of biodiversity requires conservation initiatives, habitat restoration, sustainable resource management, and the creation of protected areas. Driving revolutionary changes and encouraging a move towards more sustainable behaviors requires the coordination of international collaboration, public awareness campaigns, and policy measures.

By delving into the causes of this crisis, examining its repercussions on ecosystems and human societies, and examining case studies of biodiversity loss in various ecosystems, we hope to raise awareness of the worries and threats brought on by the loss of biodiversity. We also hope to discuss potential solutions and conservation strategies. We can all work together to create a sustainable future that puts the preservation and restoration of Earth's priceless biodiversity first by increasing awareness of the problem and realizing its seriousness (Kaltenborn, 2018).

5.2 ECOSYSTEM DIVERSITY

The diversity of the various ecosystems that exist on Earth is referred to as ecosystem diversity. Living things (plants, animals, and microorganisms) and their physical environments, such as the atmosphere, soil, water, and climate, make up ecosystems. They can include deserts, grasslands, coral reefs, and rainforests. There are distinct species, environmental factors, and ecological processes in every ecosystem. Ecosystems come in a wide variety of forms, from vast forests and grasslands to delicate coral reefs and dynamic wetlands. Numerous geographical regions, including the tropics, the poles, and both terrestrial and aquatic habitats, are home to them (Agarwal, 2015).

5.2.1 Terrestrial Ecosystems

Included in terrestrial ecosystems is the wide range of environments on earth. They range from luxuriant tropical rainforests to coniferous forests in temperate climates, and they include several kinds of forests. Examples of grasslands, which are significant terrestrial ecosystems, include savannas and prairies. Some further examples include chaparral, tundra, shrublands, and deserts. Terrestrial ecosystems are unique due to the unique mix of flora, fauna, microbes, soil types, and climatic factors that make them up.

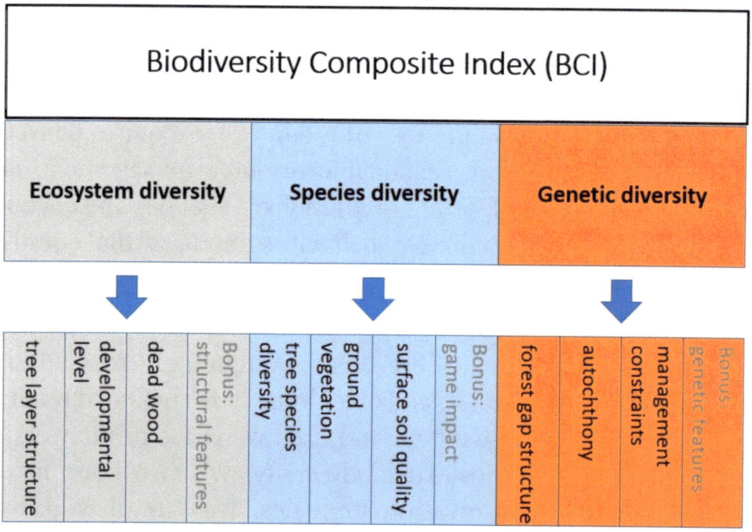

Figure 5.2: BCI index structure. The modular structure equally considers ecosystem, species, and genetic diversity with three biodiversity indicators (black font color) and one bonus indicator (gray font color), respectively. Indicators are aggregated without weighting.

Source: Ette J-S, Sallmannshofer M, Geburek T. Assessing Forest Biodiversity: A Novel Index to Consider Ecosystem, Species, and Genetic Diversity. *Forests*. 2023; 14(4):709. https://doi.org/10.3390/f14040709 [accessed 8 Jun, 2023]

5.2.2 Freshwater Ecosystems

Waterbodies including rivers, lakes, ponds, and marshes make up freshwater ecosystems. Diverse aquatic animals, such as fish, amphibians, crustaceans, and different plants adapted to live in freshwater conditions, can be found there. Water flow, temperature, nitrogen levels, and the presence of submerged or emergent vegetation all have an impact on freshwater ecosystems. They are essential in preserving water quality, controlling water flow, and assisting several biological processes (Kaltenborn, 2018).

5.2.3 Marine and Coastal Ecosystems

The immense expanse of oceans, and seas, and the dynamic contact between land and sea are all included in marine and coastal ecosystems. They consist of deep-sea ecosystems, coral reefs, kelp forests, mangrove forests, seagrass meadows, and estuaries. Astonishingly diverse marine life, from microscopic plankton to massive marine mammals, is supported by these environments. Numerous species rely on them as critical feeding grounds, nurseries, and breeding grounds. They also produce oxygen for the atmosphere and protect the coast from erosion and storms.

Each ecosystem has its collection of ecological interactions and ecological processes. They are distinguished by complex energy flows, nutrient cycling systems, and food webs. For instance, trees in a forest environment use photosynthesis to absorb sunlight, which fuels the ecosystem and provides homes for a vast array of living things. Herbivores consume the leaves, fruits, and other components of plants, while predators eat herbivores to keep the ecosystem in balance. Decomposers disintegrate organic material, redistributing nutrients to the soil and promoting resource recycling (Agarwal, 2015).

The resilience and stability of our world depend on the diversity of our ecosystems. Numerous ecosystem services offered by various ecosystems are essential for human well-being. Wetlands serve as natural filters, purifying water and enhancing its quality, while forests act as carbon dioxide sinks and releasers to help control climate. Coral reefs provide homes for a variety of marine organisms, including fish that are valuable for commerce, and shield coastlines from erosion. Each ecosystem supports the existence and survival of species and contributes to the general health of the biosphere.

Ecosystem variety is unfortunately seriously threatened by human activities including deforestation, pollution, habitat degradation, and climate change. Ecosystems are harmed and fragmented as a result of these activities, which disturbs their structure and functionality. The loss of biodiversity within ecosystems increases as they deteriorate, having an influence on the delivery of crucial ecosystem services and jeopardizing the safety of both people and animals.

The maintenance of healthy and resilient ecosystems, the preservation of ecological processes, and the sustainable use of natural resources all depend on conserving and safeguarding ecosystem variety. This can be accomplished through creating and managing protected areas, implementing sustainable land-use techniques, restoring harmed ecosystems, and incorporating ecosystem-based thinking into decision-making processes. We can work toward a future where ecosystems thrive, species flourish, and the delicate balance of life on Earth is protected by realizing the importance and value of ecosystem variety.

5.3 SPECIES DIVERSITY

The diversity of species found within a given ecosystem, geographic area, or even the entire planet is referred to as species diversity. A group of organisms that may interbreed to create fertile offspring and have similar traits is referred to as species. The number of species present as well as the relative abundance of each species are both parts of the concept of species variety. It includes the enormous variety of species that live in varied ecosystems around the world, including plants, animals, fungi, and microbes. A key element of biodiversity is species diversity, which is also essential to the health and stability of ecosystems (Kaltenborn, 2018).

5.3.1 Species Richness

The total number of various species found in a certain location or habitat is referred to as species richness. It gives an indication of the degree of diversity present in a given habitat or area. An environment with a higher species richness has a wider diversity of living things, which reflects a healthier and more robust ecosystem. For instance, a tropical rainforest can have a particularly high species diversity, with thousands of unique plant and animal species living there.

5.3.2 Evenness of Species

The relative abundance of various species within an environment is referred to as species evenness. It quantifies how equally or unequally individuals are dispersed throughout the different species that are there. Low species evenness means that some species dominate the ecosystem, which could cause imbalances and lessen total diversity. High species evenness suggests that the population sizes of different species are generally balanced (Agarwal, 2015).

5.3.3 Keystone Species

Some species are disproportionately important to maintaining the structure and performance of an ecosystem. These are the keystone species. The ecology as a whole is greatly impacted by their presence, despite the fact that they may not be the most common species. Disruptions in species interactions and environmental processes may result from the eradication of a keystone species. One such keystone species in kelp forest ecosystems is the sea otter. Its presence aids in controlling sea urchin numbers, preventing overgrazing of kelp and preserving the ecosystem's overall balance.

5.3.4 Species Used As Indicators

Species used as indicators are organisms that can offer important information about the strength and condition of an ecosystem. They can act as early warning indicators of ecosystem degradation since they are extremely sensitive to environmental changes. Scientists and conservationists may learn a lot about the condition and functioning of an ecosystem by keeping an eye on the population and behavior of indicator species. For instance, amphibians are frequently utilized as environmental quality indicators because of their porous skin, which makes them extremely vulnerable to pollutants and habitat changes.

5.3.5 Endemic Species

Endemic species can only be found in a certain geographic area and nowhere else in the globe. They only inhabit a small area and are frequently well-suited to their particular habitats. Because of their limited distribution, endemic species are particularly sensitive to challenges to conservation due to habitat loss, climate change, and other factors. It is essential to safeguard endemic species' habitats to ensure their survival and sustain the distinctive biodiversity of certain areas.

For ecosystems to be stable and resilient, species variety is crucial. Each species fills a specific ecological niche and contributes to the ecosystem's health through interactions with other species and its surroundings.

Ecological processes can be disrupted by a loss of species diversity, which could result in imbalances and eventual ecosystem collapse. Additionally, species diversity makes ecosystems resilient to environmental change because it raises the likelihood that some individuals or populations will be able to adapt (Kaltenborn, 2018).

The diversity of species is seriously threatened by human activities like habitat destruction, pollution, overexploitation, the introduction of invasive species, and climate change. These activities result in habitat degradation and fragmentation, reduced population sizes, and increased extinction rates. In addition to the loss of important ecological functions, the loss of species diversity also lessens the cultural, artistic, and ethical qualities associated with varied life forms.

The establishment of protected areas, habitat restoration, captive breeding programs, and the enforcement of laws to prevent overexploitation are just a few of the strategies used in conservation efforts to protect species diversity. Additionally, raising public awareness, promoting sustainable practices, and supporting community-based conservation initiatives are essential for safeguarding species diversity and ensuring the long-term survival of Earth's diverse array of life (Agarwal, 2015).

To sum up, species diversity is an essential part of biodiversity and plays a role in the resilience, stability, and efficiency of ecosystems. The preservation and conservation of species diversity are vital for maintaining ecosystem health, supporting human well-being, and preserving the beauties of our natural world.

5.4 GENETIC DIVERSITY

The variance in the genetic makeup of individuals within a species is referred to as genetic diversity. Genetic variations that result from mutations, recombination during reproduction, and gene flow between populations are the cause of it. Genetic variation is essential for a species to adapt and be resilient to environmental changes because it gives natural selection the starting point on which to work. Additionally, it helps to keep ecosystems viable and stable overall. Genetic variation emerges through a combination of events, including mutations, genetic recombination during reproduction, and gene flow across populations.

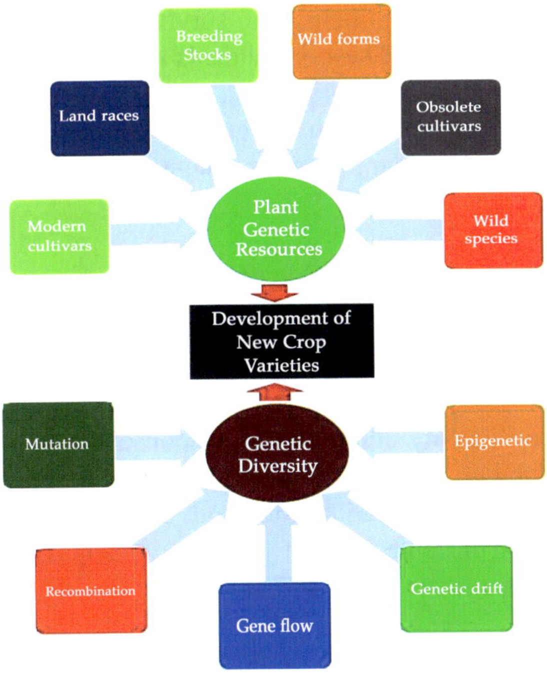

Figure 5.3: Different sources of genetic diversity and their potential utilization in the development of new crop varieties.

Source: Salgotra, R. K., & Chauhan, B. S. (2023). Genetic Diversity, Conservation, and Utilization of Plant Genetic Resources. *Genes*, 14(1), 174. MDPI AG. Retrieved from http://dx.doi.org/10.3390/genes14010174

5.4.1 Relevance of Genetic Variation

Genetic variation is of the highest relevance for the long-term survival and adaptability of species. It provides the raw material for natural selection, enabling populations to evolve and adapt to changing environmental conditions. A species' resistance to diseases, pests, and other stresses is increased by greater genetic diversity within the species. Additionally, it increases the likelihood that people will possess the genetic characteristics required for surviving and procreating in a changing environment (Kaltenborn, 2018).

5.4.2 Evolution and Adaptation

The occurrence of adaptation and evolution depends on genetic variation. Populations with greater genetic diversity are more likely to have individuals with features that help them survive and procreate in the face of environmental changes. These individuals can transmit their favorable characteristics to succeeding generations, aiding in the species' ability to adapt and evolve. Populations with low genetic diversity, on the other hand, might find it difficult to adapt to environmental challenges, making them more susceptible to extinction.

5.4.3 Genetic Variation and Ecosystem Function

Genetic variation within a species is necessary for the survival of the species and also benefits the health of an ecosystem as a whole. Intraspecific genetic diversity may have an impact on interactions and ecological processes. For instance, genetic diversity and a plant's response to its environment can affect how productive, insect-resistant, and resilient a population of plants is. The ability of animal populations to reproduce, their resistance to disease, and the preservation of stable predator-prey relationships can all be impacted by genetic diversity (Agarwal, N. K. 2015).

5.4.4 Genetic Diversity and Conservation

If species and ecosystems are to remain long-term viable, genetic diversity must be maintained. Numerous factors, including habitat loss, population bottlenecks, habitat fragmentation, inbreeding, and human actions that obstruct normal gene flow, can lead to genetic diversity loss. Populations become less capable of adapting as genetic variation declines, making them more sensitive to environmental changes.

The creation of protected areas that include representative populations and habitats, the administration of captive breeding programs, and the reintroduction of genetically varied individuals into declining populations are all conservation measures aimed at maintaining genetic diversity. Additionally, sustaining genetic variation depends on protecting natural migration routes and fostering gene exchange between dispersed groups.

5.4.5 Genetic Diversity and Human Well-Being

Genetic diversity is essential for both human well-being and the preservation of wild species. Agriculture, medicine, and industry have long utilized genetic resources from a variety of species. The basis for crop improvement, the invention of novel medications, and the development of resistant and fruitful livestock breeds is genetic features present in many populations and species. Maintaining genetic diversity assures that genetic resources will be accessible for upcoming discoveries and advantages (Kaltenborn, 2018).

In conclusion, genetic variation is essential for species adaptation, evolution, and long-term survival. It improves population and ecological resilience, increases their capacity to adapt to environmental changes, and promotes ecosystem health. The preservation of genetic variation within and between populations must be a priority of conservation efforts in addition to the protection of species and ecosystems. We can assist sustain the inherent value of species, aid ecological processes, and ensure the potential advantages that natural genetic resources offer to human society by preserving genetic variety.

These three elements of biodiversity are intricately entwined and rely on one another. A wide variety of species, each with a distinct genetic makeup, make up ecosystems. Genetic diversity within species plays a role in the overall adaptation and resilience of ecosystems. The intricate web of life can flourish because ecosystems can support species existence and survival.

To keep our world stable and healthy, biodiversity is essential. Numerous crucial functions are provided by ecosystems, such as temperature regulation, air and water purification, nutrient cycling, plant pollination, and the provision of food and resources for human populations. These ecosystem services are supported by the diversity of species and genetic variation within species, which also supports the overall health and productivity of ecosystems.

However, the disappearance of biodiversity has raised urgent issues. Global biodiversity has significantly decreased as a result of human actions such as habitat destruction, overuse of resources, pollution, climate change, and the introduction of exotic species. The sustainability of human societies, the survival of animals, and ecosystems are all gravely threatened by this loss. For the protection of our natural heritage and the welfare of upcoming generations, an understanding of and commitment to biodiversity is essential. By appreciating the importance and worth of biodiversity, we can work toward sustainable practices, safeguard species and their habitats, repair damaged ecosystems, and encourage the peaceful coexistence of all life forms on Earth (Agarwal, 2015).

5.5 CAUSES OF BIODIVERSITY LOSS

The loss of biodiversity has dramatically accelerated in recent decades, primarily as a result of human activity. Effects of these activities include significant habitat damage, excessive resource consumption, pollution, climate change, and the introduction of invasive species. Understanding the reasons for biodiversity loss is crucial for addressing and reducing these problems.

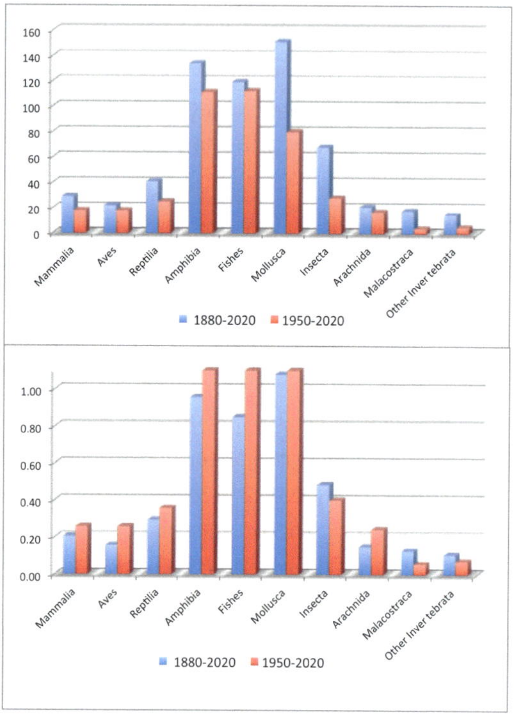

Figure 5.4: Comparison between the number of species extinct during the last 140 and 70 years in the major taxonomical groups of animals.

Source: Palombo MR. Thinking about the Biodiversity Loss in This Changing World. *Geosciences*. 2021; 11(9):370. https://doi.org/10.3390/geosciences11090370

The following are a few of the main causes:

- *Habitat Damage and Fragmentation*: One of the main reasons for the decline in biodiversity is habitat damage. Numerous species lose vital habitats as a result of the change of natural ecosystems, such as forests, wetlands, and grasslands, into agricultural land, urban areas, and infrastructure development. Because it destroys the intricate and varied ecosystems found in forests and replaces them with simpler landscapes, deforestation is one of the main causes of biodiversity loss. The division of huge habitats into smaller, isolated pieces, known as habitat fragmentation, also disturbs biological processes and restricts the capacity of species to move and locate suitable habitats (Kaltenborn 2018).

- *Overexploitation of Natural Resources*: Overexploitation happens when a species is harvested or extracted at a faster rate than it can naturally regenerate. Unsustainable plant and animal harvesting, fishing, logging, and hunting can result in the decline and even extinction of certain species. Examples include killing charismatic animals like rhinos and elephants for their horns and ivory as well as overfishing popular marine species like bluefin tuna. Overexploitation affects not only the targeted species but also other species that depend on them by upsetting food webs, dwindling populations, and altering ecosystems.
- *Pollution*: Pollution, including chemical and physical pollution, poses a major danger to biodiversity. Air, water, and soil can be contaminated by industrial pollution, agricultural runoff, incorrect waste disposal, and the discharge of harmful compounds into ecosystems. Pollution can cause direct injury to living things, destabilize entire ecosystems, and make it harder for many species to find suitable habitats. For instance, pesticides and fertilizers can contaminate water, which can harm fish, amphibians, and other species that depend on water.
- *Climate Change*: A global phenomenon, climate change is caused by an increase in greenhouse gas emissions, mostly due to human activity. Climate change has a significant influence on biodiversity due to its rising temperatures, changed rainfall patterns, and extreme weather events. Many species find it difficult to migrate or adapt quickly enough to keep up with these swift changes, which cause range shifts, phenological changes, and increased disease susceptibility. The availability of suitable habitats is also impacted by climate change, which also has the potential to upset important ecological interactions like those between pollinators and plants (Agarwal, 2015).
- *Introduction of Invasive Species*: The native biodiversity might suffer when invasive or non-native species are introduced to habitats. Natural rivals or predators are frequently absent from invasive species' new settings, allowing them to outcompete native species for resources and alter ecological processes. They can damage habitats, change food chains, and bring about the

decrease or extinction of native species. Invasive species like kudzu or animals like the brown tree snake in Guam are two examples.

- *Human Expansion and Population Growth*: The expanding human population and its impact on natural regions put extra strain on biodiversity. As a result of increasing agricultural practices, land conversion, and resource extraction due to rising food, water, energy, and resource demands, habitats are being degraded and destroyed. Infrastructure expansion and urbanization obstruct migration and fragment habitats, isolating populations.
- *Agricultural Intensification*: Agriculture is becoming more intensive as a result of rising food production demands, which has a detrimental effect on biodiversity. The loss of biodiversity is a result of large-scale monoculture farming, the use of synthetic fertilizers and pesticides, and the conversion of various natural habitats into agricultural land. These methods have the potential to destroy natural habitats, reduce soil fertility, taint water supplies, and eliminate pollinators and beneficial insects (Kaltenborn, 2018).
- *Land Use Change*: As a result of human activities including urbanization, infrastructure construction, and industrial growth, natural landscapes are transformed into artificial settings. The destruction of ecosystems, disruption of ecological connectedness, and reduction of suitable habitats for many species result from the clearing of land for roads, buildings, and other infrastructure. Additionally, harming biodiversity, land use change causes the loss of ecosystem services like soil fertility, climate regulation, and water regulation.
- *Unsustainable Forestry Practices*: Negative logging methods like clear-cutting and selective logging can seriously harm forest ecosystems. Clear-cutting, which results in habitat loss and soil erosion, includes the removal of all the trees in a specified region. Even though selective logging is more focused, it can still damage the ecosystems of forests by eliminating particular tree species and changing the composition and structure of the forest. Unsustainable forestry practices can result in the destruction of forest ecosystems, the loss of biodiversity, and the interruption of vital ecological processes.

- *Genetic Manipulation and Genetic Pollution*: Both genetic modification and the introduction of genetically modified organisms (GMOs) into the environment have the potential to have unforeseen effects on biodiversity. Potential interbreeding between GMOs and wild cousins could contaminate genetic diversity and destroy the genetic integrity of wild populations. The genetic diversity and adaptability of native species may be negatively impacted over the long term by this.
- *Wildlife Trade*: The demand for unusual pets, traditional remedies, luxury items, and ornamental plants is driving the illegal wildlife trade, which poses a danger to biodiversity. Elephants, tigers, and rhinos are just a few of the threatened species that are being driven closer to extinction by poaching and trafficking for their live specimens or body parts. A further factor in the decline of natural populations is the unsustainable collection and trading of rare flora and animals for horticulture and the pet industries (Agarwal, 2015).
- *Lack of Environmental Consciousness and Education*: The loss of biodiversity can be attributed to a lack of environmental consciousness and understanding of the value of biodiversity. People may engage in unsustainable behaviors, contribute to the destruction of habitats, or support activities that affect biodiversity when they are unaware of the value of biodiversity and the detrimental effects of their behavior. Promoting conservation efforts and instilling a sense of responsibility for the natural environment depends on education and creating knowledge about the value of biodiversity and sustainable practices (Kaltenborn, 2018).

It takes coordinated efforts at many different levels, including legislative interventions, sustainable land and resource management practices, conservation programs, and public awareness and education, to address the causes of biodiversity loss. The key to conserving biodiversity and ensuring the health of ecosystems and human societies is to protect and restore habitats, implement sustainable farming and fishing methods, reduce pollution, and promote the sustainable use of resources.

5.6 CONSEQUENCES OF BIODIVERSITY LOSS

Ecosystems, economics, and human welfare are all negatively impacted by the loss of biodiversity. Direct and indirect effects of biodiversity loss include effects on ecological functions, ecosystem services, and the long-term viability of our planet. The following are some important effects of biodiversity loss:

- ***Ecosystem Function Disruption***: The rich network of interconnections within ecosystems is disrupted by biodiversity loss. Each species has a special responsibility for preserving ecosystem functions like the cycling of nutrients, pollination, seed dissemination, and pest control. The elimination of species due to extinction or population decline can cause imbalances and disruptions in these processes, endangering the ecosystems' general stability and ability to function. Reduced production, higher susceptibility to infectious illnesses and invasive species, and decreased resistance to environmental changes can all arise from this.
- ***Reduced Adaptability and Resilience***: Ecosystems with a variety of life have a built-in capacity to withstand shocks and bounce back. High biodiversity ensures that certain species have the genetic traits and adaptations they need to live and reproduce in a changing environment. As biodiversity declines, ecosystems become less resilient to environmental disturbances such as climate change, habitat loss, and pollution. Reduced resilience renders ecosystems more susceptible to further degradation and raises the chance of ecosystem collapse.
- ***Loss of Ecosystem Services***: Ecosystem services cover supplying (food, water, and timber), regulating (climate control, water purification), cultural (recreation, spiritual, and aesthetic qualities), and supporting (soil formation, and nutrient cycling) functions. The advantages that come from nature to humans are called ecosystem services. These services are directly impacted by biodiversity loss. For instance, the loss of wetlands reduces the ability to purify water, while the fall of pollinators undermines the production of crops. Reduced biodiversity jeopardizes ecosystems' ability to deliver these crucial functions, putting human livelihoods and well-being in danger (Agarwal, 2015).
- ***Increased Exposure to Climate Change***: The loss of biodiversity and climate change are linked. Ecosystems are less equipped to withstand the effects of climate change as biodiversity disappears. Ecosystems that are rich in biodiversity can store

and sequester carbon, control regional temperatures, and act as natural barriers against extreme weather. These natural climate regulating mechanisms are diminished by biodiversity loss, which exacerbates the effects of climate change. In addition, altered temperature and precipitation patterns brought on by climate change may put further strain on already fragile and fragmented ecosystems, speeding up the loss of species (Kaltenborn, 2018).

- ***Reduced Access to Genetic and Medical Resources***: Many species, particularly those from the plant and animal kingdoms, have unique genetic traits and biochemical components that have been demonstrated to be advantageous for human health and well-being. The likelihood of discovering new medicines and genetic resources for biotechnology, agricultural, and medical developments decreases as biodiversity declines. It limits our ability to develop new treatments for illnesses that are on the rise and to improve agricultural productivity.
- ***Cultural and Aesthetic Impacts***: Loss of biodiversity has an impact on culture and aesthetics, as well. Indigenous and local groups usually rely on a variety of ecosystems for their cultural identity, practices, and way of life, and frequently have strong spiritual and cultural ties to their natural environment. These cultural customs and values are disturbed when species, landscapes, and ecosystems vanish. Furthermore, the loss of biodiversity affects our emotional and psychological well-being by reducing the aesthetic richness and diversity of our natural environment.
- ***Risk to Global Food Security***: The loss of biodiversity poses a serious threat to the world's ability to feed itself. Pollination services necessary for the growth of many fruits, vegetables, and nuts are being reduced as a result of the reduction of pollinators like bees and butterflies. Additionally, crop plants and livestock varieties are becoming more susceptible to pests, illnesses, and environmental pressures, which pose a danger to agricultural productivity. For sustainable food production and global food security, it is essential to have robust and varied agricultural systems.
- ***Increased Rates of Extinction***: Increased extinction rates are directly correlated with biodiversity loss. Unique evolutionary

lineages and genetic diversity are permanently lost with the extinction of species. The delicate web of life is upset when a species goes extinct, and this can have a domino impact on entire ecosystems. Keystone species loss can result in substantial disturbances and perhaps the collapse of ecological communities because they have disproportionately big effects on their ecosystems.

- ***A Decline in Ecotourism and Nature-based Recreation***: The tourism sector, particularly ecotourism and nature-based recreation, can be impacted by biodiversity loss. Natural places draw a lot of people who want to see the various species, pristine landscapes, and distinctive ecosystems that they contain. The loss of biodiversity reduces the allure of these locations, resulting in fewer travelers and the corresponding economic benefits. Local economies may be significantly impacted by this, particularly in areas where nature-based tourism is prevalent (Agarwal, 2015).
- ***Social and Economic Inequality***: Social and economic inequality is exacerbated by biodiversity loss, which frequently has a disproportionately negative impact on vulnerable groups. Indigenous peoples and local communities depend on biodiversity for their survival, cultural traditions, and traditional knowledge. These groups frequently have intimate interactions with the ecosystems around them. These people experience food insecurity, loss of livelihoods, and eroding cultural traditions as a result of declining biodiversity, which enlarges the gap between various social groups (Kaltenborn, 2018).
- ***Increased Danger of Zoonotic Diseases***: The danger of zoonotic diseases, which are infectious illnesses spread between animals and people, can increase as a result of biodiversity loss and the destruction of natural ecosystems. Wildlife species may come into closer contact with human populations when natural habitats are disrupted or destroyed, raising the risk of disease transmission. Diseases like Ebola, SARS, and COVID-19 can originate and spread as a result of the loss of biodiversity upsetting the harmony of the body's natural disease control mechanisms.
- ***Loss of Ethical and Intrinsic Values***: The extinction of species calls into question our moral obligation to safeguard other species and presents ethical issues. Regardless matter how valuable a species is to humans; it still has intrinsic value and a right to live.

The richness and diversity of life on Earth, as well as our moral and ethical ties to it, are being reduced through biodiversity loss. The significance of maintaining ecosystem integrity and appreciating the inherent value of all species are also key components of preserving biodiversity, in addition to the benefits it offers on a practical level (Kaltenborn, 2018).

Action that is swift and well-coordinated is required to combat biodiversity loss and its consequences. Through conservation initiatives, sustainable land and resource management strategies, the establishment of protected areas, habitat restoration, and international cooperation, it is necessary to safeguard the crucial contributions that biodiversity contributes to ecosystems, economies, and human well-being.

5.7 CONSERVATION AND RESTORATION EFFORTS

The extinction of species calls into question our moral obligation to safeguard other species and presents ethical issues. Regardless matter how valuable a species is to humans; it still has intrinsic value and a right to live. The richness and diversity of life on Earth, as well as our moral and ethical ties to it, are being reduced through biodiversity loss (Agarwal, 2015).

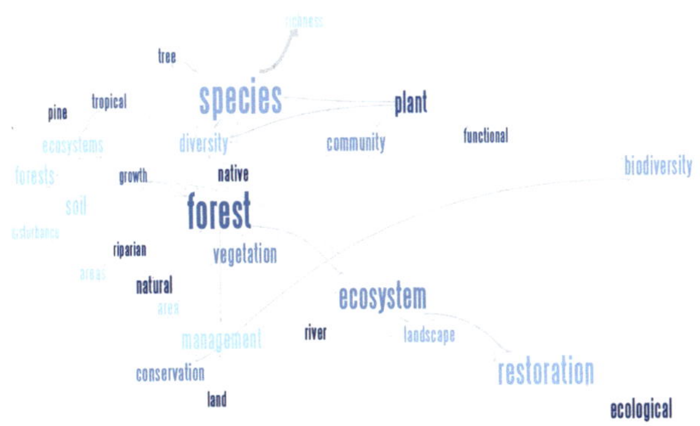

Figure 5.5: Related concepts in the scientific literature on forest restoration, biodiversity and ecosystem functioning.

Source: Aerts, R., Honnay, O. Forest restoration, biodiversity and ecosystem functioning. *BMC Ecol* 11, 29 (2011). https://doi.org/10.1186/1472-6785-11-29

The significance of maintaining ecosystem integrity and appreciating the inherent value of all species are also key components of preserving biodiversity, in addition to the benefits it offers on a practical level.

- *Protected Places*: Preserving biodiversity requires protected places, such as national parks, wildlife refuges, and marine sanctuaries. These places serve as habitats for a variety of species, enable ecosystem restoration, and sustain vital biological processes. The creation and successful management of protected areas, in partnership with regional stakeholders and local populations, contribute to the preservation of ecosystems, biodiversity, and ecosystem services. To allow for the migration of species and to permit their adaptability to changing settings, the expansion and connection of protected areas are also crucial.
- *Habitat Restoration and Reforestation*: Reforestation and habitat restoration are important for fostering the return of biodiversity. Reforestation, wetland rehabilitation, and the eradication of invasive species are some examples of habitat restoration techniques. Degraded habitats can restore their ecological functions and promote the return of varied plant and animal groups by reintroducing natural vegetation and re-creating habitat conditions that are favorable for species. The fight against deforestation and its effects on biodiversity, climate change, and soil erosion is especially important (Kaltenborn, 2018).
- *Sustainable Land and Resource Management*: It is crucial to use sustainable land and resource management techniques to protect biodiversity. This involves advocating for sustainable agricultural practices, such as agroforestry and organic farming, which use fewer toxic pesticides and fertilizers, preserve the health of the soil, and encourage farming methods that are favorable to biodiversity. To maintain habitat integrity and stop additional biodiversity loss, sustainable forestry practices, such as selective logging and the protection of old-growth forests, are used. To safeguard vulnerable species and maintain healthy marine ecosystems, sustainable fisheries management is essential (Agarwal, 2015).
- *Conservation of Species*: To save threatened species and stop their extinction, focused efforts are required. The protection of vital ecosystems, the captive breeding and reintroduction of endangered species, habitat restoration tailored to their needs, and the lowering of threats like poaching and the illegal wildlife trade are all examples of conservation efforts. Furthermore,

protecting keystone and umbrella species—species that, when protected, indirectly benefit other species—contributes to the general preservation of biodiversity and the health of ecosystems.

- *Community-based Conservation*: Long-term effectiveness in conservation requires including local communities. Fostering a sense of ownership and care toward biodiversity involves involving communities in decision-making, acknowledging and honoring their traditional knowledge and traditions, and offering financial incentives for conservation efforts. The empowerment of local communities to actively participate in conservation efforts through collaborative management techniques, such as community-based conservation programs and co-management of protected areas, ensures the sustainable use of resources and the preservation of biodiversity.
- *International Cooperation and Policy Interventions*: International cooperation and policy changes are crucial for halting the loss of biodiversity since many species and habitats cross national boundaries. International agreements like the Convention on Biological Diversity (CBD) provide a framework for collaboration between nations so that biodiversity may be preserved, sustainable development can be advanced, and resources and information can be shared. Policy interventions, such as the development and enforcement of environmental rules, the inclusion of biodiversity considerations in land-use planning, and the promotion of sustainable practices, are essential to mainstreaming biodiversity conservation throughout numerous sectors.
- *Public Awareness and Education*: Promoting environmental education and increasing public awareness are essential for developing a culture of biodiversity protection. Campaigns, educational programs, and nature-based tourism are examples of public engagement activities that aid in educating the public about the importance of biodiversity and the effects of its loss. Public awareness and education support broader conservation efforts by encouraging a sense of responsibility, supporting sustainable lifestyle choices, and motivating action (Kaltenborn, 2018).

Governments, organizations, communities, and people must work together and be committed to conservation and restoration activities. We can significantly advance biodiversity conservation and restoration, ecosystem protection, and ensure a sustainable future for future generations by combining these techniques and working together.

5.8 FUTURE OUTLOOK

Both difficult and exciting prospects exist for biodiversity preservation. Even as threats to biodiversity continue to worsen, it is becoming increasingly clear that rapid action is required and that protecting and restoring ecosystems is important. Here are some key components of the forecast for future biodiversity protection:

- *Heightened Awareness and Global Commitment*: The value of biodiversity and the consequences of losing it are now better understood by the public. Governments, businesses, and individuals are all becoming more conscious of the need to prioritize and factor biodiversity conservation into choices. Worldwide agreements and initiatives like the Post-2020 Global Biodiversity Framework and the Sustainable Development Goals of the United Nations provide a framework for worldwide commitment to and action toward biodiversity protection (Agarwal, 2015).
- *Technology and Scientific Research Advancements*: These developments present great chances for biodiversity preservation. Species distributions, population dynamics, and the health of ecosystems can all be better understood with the help of DNA barcoding, remote sensing, and other monitoring methods. Emerging technologies can also help with data analysis, projecting changes in biodiversity, and identifying successful conservation methods. Examples include artificial intelligence and machine learning. These scientific developments support creative conservation strategies and evidence-based decision-making.
- *Integration of Biodiversity into Development*: For a sustainable future, it is essential to include biodiversity conservation in development sectors. It is becoming more popular to acknowledge the connections between biodiversity, climate change, poverty reduction, food security, and other development objectives. To reduce detrimental effects on biodiversity and enhance the beneficial contributions of these sectorsto conservation, efforts should be made to mainstream biodiversity considerations into

industries like agriculture, forestry, fisheries, infrastructure development, and tourism (Kaltenborn, 2018).

- *Approaches at the Landscape Level*: Beyond protected areas, the necessity of landscape-level methods for biodiversity conservation is becoming more and more apparent. This entails maintaining connectivity and preserving habitats across entire landscapes, including both protected and unprotected areas. Key elements of such methods include multi-stakeholder alliances, collaborative governance, and integrated landscape management strategies. Landscape-level conservation encourages sustainable land use methods, increases ecological resilience, and makes it easier for species to move around (Agarwal, 2015).

Chapter

6 BURDENS, CAUSES, AND RESPONSE OF CHRONIC ENVIRONMENTAL DISEASE

CONTENTS

6.1 INTRODUCTION

People's health and well-being are significantly impacted by chronic environmental illnesses, which can result in a variety of health problems affecting different bodily systems. Effective early detection, diagnosis, as well as therapy of many diseases, depend on a thorough understanding of the unique circumstances connected to them. Continuous health issues that are brought on by or made worse by exposures to environmental variables are referred to by the term chronic environmental illnesses. In-depth studies of the burdens, causes, and reactions to chronic environmental illnesses are provided in this chapter (Balmes, 2010).

6.2 CHRONIC ENVIRONMENTAL ILLNESS BURDENS

6.2.1 Health Effects

People with chronic environmental disorders frequently struggle with respiratory disorders. Respiratory issues can result from exposure to air contaminants particulate matter, ozone, nitrogen dioxide as well as volatile organic compounds. These consist of:

- **Asthma**: The hallmark signs of asthma, a long-lasting respiratory disorder marked by inflammation with airway constriction, include wheezing, chest tightness, coughing, and shortness of breath. Although genetics and personal risk factors play a factor in the beginning and course of asthma, environmental factors, particularly pollutants in the air, can have a considerable impact on the condition. Particulate matter, ozone, nitrogen dioxide, as well as volatile organic compounds are examples of air pollutants that have been shown to initiate or exacerbate asthma symptoms.

  Tiny airborne particles suspended in the atmosphere are referred to as particulate matter. According to studies, being exposed to PM, especially particulates with a diameter of 2.5 micrometers or less, is linked to worsening asthma symptoms and damaged lungs. These particles can enter the respiratory system deeply, inflaming the airways and bringing on asthma episodes.

  Nitrogen dioxide is a gas that is produced when fossil fuels are burned, primarily in cars and power plants. The emergence of asthma in children and the worsening of asthma symptoms in people who already have the condition have both been related to prolonged exposure to NO2. This gas can dry out the airways, causing inflammation or bronchial hyperresponsiveness (Zhong, 2016)

Pollutant ozone is produced when sunlight reacts with nitrogen oxides and volatile organic molecules. Studies have repeatedly demonstrated that being exposed to O3 is linked to an increase in asthma symptoms, ER visits, and hospital admissions. NO3 can lead to oxidative stress including airway inflammation, which can start asthma episodes (Balmes, 2010).

Volatile organic compounds (VOCs) are released from many sources, such as everyday items, construction materials, and industrial operations. Formaldehyde, benzene, and toluene are a few VOCs that have been connected to the onset and worsening of asthma. In those who are vulnerable, these substances may damage the airways, causing swelling and respiratory problems.

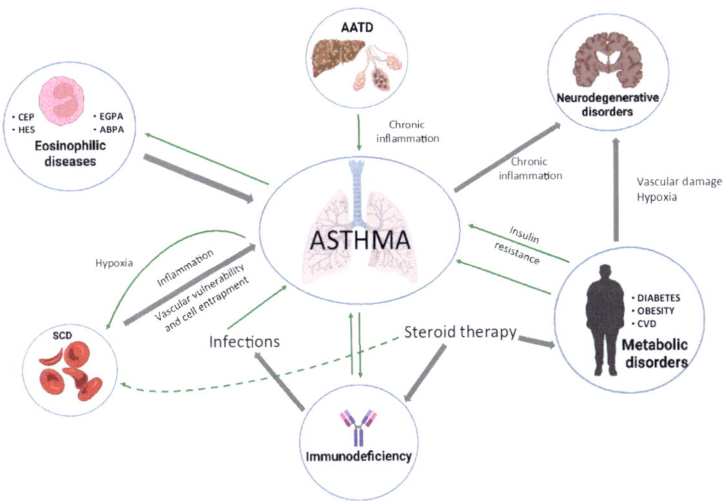

Figure 6.1: Overview of asthma.

Source: Maule M, Olivieri B, Guarnieri G, De Franceschi L, Martinelli N, Vaia R, Argentino G, Vianello A, Senna G, Caminati M. Hidden Comorbidities in Asthma: A Perspective for a Personalized Approach. *Journal of Clinical Medicine*. 2023; 12(6):2294. https://doi.org/10.3390/jcm12062294 [accessed 9 Jun, 2023]

- **Chronic Obstructive Pulmonary Disease (COPD)**: Chronic obstructive pulmonary disease has repeatedly been linked to prolonged exposure to air pollution, especially fine particulate matter. Chronic obstructive pulmonary disease is marked by continuous airflow restriction, breathing difficulties, and increased susceptibility to respiratory infections. Let's examine the connection between PM2.5, a kind of air pollutant, and COPD in greater depth (Zhong, 2016).

Several investigations in the scientific community have shown a direct link between chronic exposure to PM2.5 and both the development and progression of COPD. PM2.5 refers to tiny particles having a diameter of 2.5 micrometers or less that can enter the lower airways including the alveoli of the respiratory system.

The risk of getting COPD has repeatedly been related to prolonged exposure to PM2.5. Higher concentrations of PM2.5 have been scientifically linked to a greater frequency of COPD in both urban and rural populations, according to epidemiological research. People who reside in locations with high concentrations of PM2.5 are more likely to acquire COPD, especially if they are exposed for a lengthy time (Balmes, 2010).

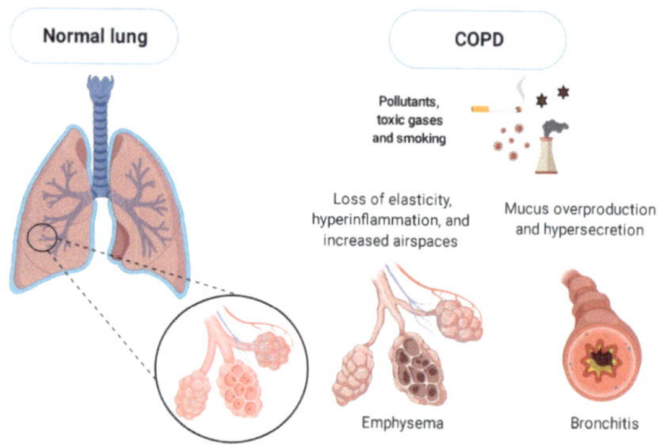

Figure 6.2: Chronic obstructive pulmonary disease (COPD).

Source: Rodrigues SdO, Cunha CMCd, Soares GMV, Silva PL, Silva AR, Gonçalves-de-Albuquerque CF. Mechanisms, Pathophysiology and Currently Proposed Treatments of Chronic Obstructive Pulmonary Disease. *Pharmaceuticals*. 2021; 14(10):979. https://doi.org/10.3390/ph14100979 [accessed 9 Jun, 2023]

- **Cardiovascular Diseases**: Cardiovascular disorders, especially ischemic heart disease, as well as stroke, are significantly influenced by environmental variables, including air pollution.

Numerous important conclusions on the relationship between long-term environmental exposures with cardiovascular illnesses have been drawn from significant research.

Ischemic heart disease is a disorder when there is insufficient blood flow to the heart's muscle cells as a result of coronary artery constriction or blockage. The risk of ischemic heart disease along with cardiovascular death has continuously been elevated with extended exposure to air pollution.

Numerous studies have shown a substantial correlation between the risk of ischemic heart disease and long-term exposure to fine particulate matter (PM2.5). PM2.5 particles are tiny enough to penetrate the circulation to cause systemic inflammation, and endothelial dysfunction increasing oxidative stress. The fundamental cause of ischemic heart disease, atherosclerosis, develops and advances as a result of these events.

Additionally, linked to a higher risk of developing ischemic heart disease is prolonged exposure to nitrogen dioxide (NO2), which is typically released during combustion activities. NO2 can cause oxidative stress, inflammation, and vascular dysfunction, which can result in atherosclerosis formation and progression.

Other air pollutants, such as sulfur dioxide (SO2) and ozone (O3), have also been linked to the onset of ischemic heart disease, albeit the data is less conclusive than that for PM2.5 as well as NO2. When the supply of blood to the brain is disrupted, a stroke ensues, which causes neurological damage. Chronic exposure to pollutants in the air has been associated with a higher risk of stroke, especially ischemic stroke, which makes up the majority of stroke occurrences (Balmes, 2010).

Ozone is an oxidant gas produced by photochemical reactions and has also been connected to a higher risk of stroke. Atherosclerosis and stroke may occur as a result of inflammation, oxidative stress , and vascular damage brought on by ozone exposure (Zhong, 2016). Nitrogen dioxide (NO_2) and the risk of stroke may be related, according to certain research, albeit the evidence is less strong than that for PM2.5 and ozone. Although more study is required to conclusively demonstrate these associations, exposure to other air pollutants, such as carbon monoxide (CO) and volatile organic compounds (VOCs), may potentially increase the risk of stroke.

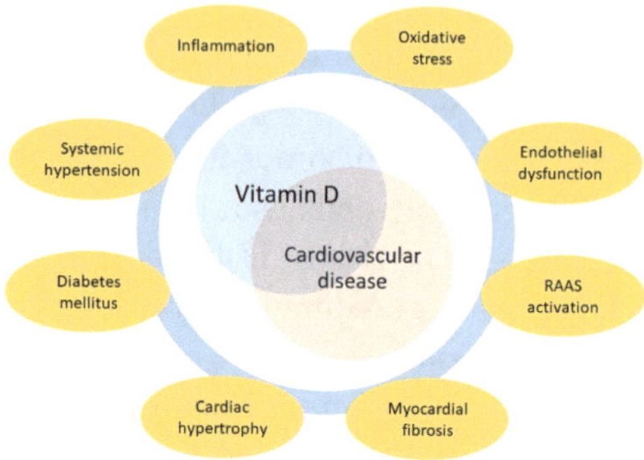

Figure 6.3: Major aspects of cardiovascular diseases.

Source: Cosentino N, Campodonico J, Milazzo V, De Metrio M, Brambilla M, Camera M, Marenzi G. Vitamin D and Cardiovascular Disease: Current Evidence and Future Perspectives. *Nutrients*. 2021; 13(10):3603. https://doi.org/10.3390/nu13103603 [accessed 9 Jun, 2023]

- **Cancer**: Numerous cancers, including bladder and lung cancer, have been discovered to be influenced by long-term environmental exposures. Numerous studies on the link between environmental variables and cancer risk have shed light on the negative effects of long-term exposures on human health.

  Lung cancer risk is significantly enhanced by prolonged exposure to air pollution. Particulate matter and certain gasses have been identified as key lung cancer risk factors among the numerous air contaminants. Numerous studies have consistently linked exposure to PM with a higher risk of developing lung cancer. Tiny particles known as PM are dispersed in the air and can include many harmful substances, including heavy metals, polycyclic aromatic hydrocarbons (PAHs), and carcinogens. When inhaled, these particles can penetrate deep into the lungs, leading to chronic inflammation, DNA damage, and the initiation of carcinogenesis.

  The risk of lung cancer can also be increased by certain airborne gases. For instance, prolonged exposure to nitrogen oxides and sulfur oxides from car exhaust and industrial pollutants has been linked to a higher prevalence of lung cancer. Secondary pollutants, such as nitrous acid and sulfuric acid, which may have cancerous consequences for the respiratory system, can be created when these gases mix with other air pollutants (Zhong, 2016).

Bladder cancer risk has been linked to exposure to environmental chemicals. This connection is influenced by several occupational as well as environmental factors (Balmes, 2010). Certain sectors expose workers to substances including aromatic amines, polycyclic aromatic hydrocarbons, and aromatic hydrocarbons that have been linked to bladder cancer. Due to occupational exposure to these compounds, individuals working in sectors including rubber manufacture, dye production, and chemical manufacturing may have an increased risk of bladder cancer.

Bladder cancer risk has been linked to drinking water pollution with certain chemicals, such as arsenic and chlorination byproducts. People exposed to high quantities of arsenic in their water-based sources have higher incidences of bladder cancer. Arsenic is a particularly well-known bladder carcinogen. A substantial risk factor for bladder cancer is cigarette smoking. Aromatic amines and polycyclic aromatic hydrocarbons, two compounds found in tobacco smoke, can enter the circulation and be eliminated in the urine, coming into contact with the bladder lining and raising the risk of bladder cancer.

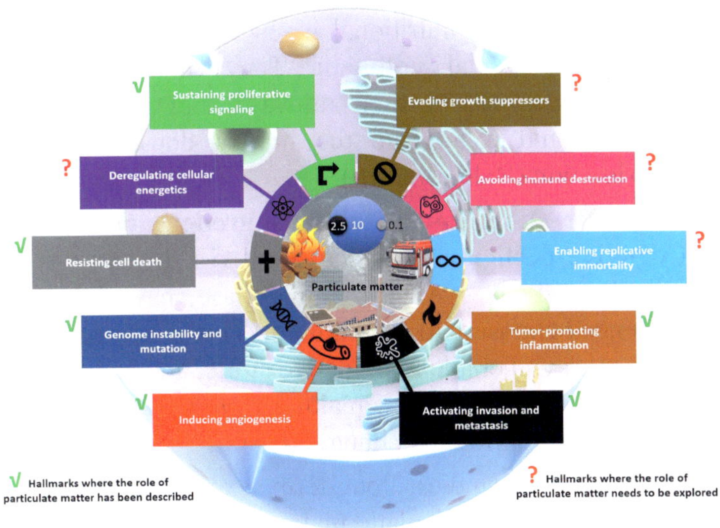

Figure 6.4: Particulate matter and its role in cancer hallmarks.

Source: Santibáñez-Andrade M, Chirino YI, González-Ramírez I, Sánchez-Pérez Y, García-Cuellar CM. Deciphering the Code between Air Pollution and Disease: The Effect of Particulate Matter on Cancer Hallmarks. *International Journal of Molecular Sciences*. 2020; 21(1):136. https://doi.org/10.3390/ijms21010136 [accessed 9 Jun, 2023]

- **Neurological Disorders**: Chronic environmental exposures can harm the nervous system, increasing the likelihood of cognitive decline and neurodegenerative illnesses like Alzheimer's disease as well as neurological conditions like Parkinson's disease and other neurological disorders. Parkinson's disease is a neurodegenerative condition marked by the loss of dopamine-producing brain cells, which causes tremors, stiffness, and movement problems. Environmental variables, such as exposure to certain pesticides and chemicals, have been cited as potential risk factors for Parkinson's disease even though the illness's specific origins are not entirely understood.

  Numerous researches have revealed a connection between pesticide exposure and an elevated risk of Parkinson's disease. An increased chance of developing Parkinson's disease has been linked to exposure to heavy metals like manganese and lead as well as organochlorine pesticides like chlorpyrifos and dieldrin. These substances can enter the body through a number of different channels, such as nutritional ingestion, occupational exposure, and environmental pollution. They might impede cellular functions, oxidative stress pathways, and protein aggregation, which would exacerbate the brain's dopaminergic neurons' deterioration (Balmes, 2010).

- **Cognitive Impairment and Neurodegenerative Diseases**: Cognitive decline and an increased risk of neurodegenerative disorders, such as Alzheimer's disease, have been related to prolonged exposure to air pollution. The brain consumes a lot of oxygen, is vulnerable to oxidative stress, and is directly exposed to airborne contaminants, making it vulnerable to the negative effects of air pollution (Zhong, 2016).

  According to studies, long-term exposure to air pollution, especially fine particulate matter (PM2.5) and contaminants from traffic, is linked to cognitive impairment and a higher chance of acquiring neurodegenerative disorders. By causing oxidative stress, systemic inflammation, and the production of pro-inflammatory cytokines, breathing in small particles has the potential to accelerate neurodegenerative and neuroinflammatory processes in the brain. The development of beta-amyloid plaques as well as neurofibrillary tangles, the defining pathology of Alzheimer's disease, could be facilitated by these consequences.

Through the olfactory route, air pollutants can directly reach the brain, or they can have an indirect impact on the brain by causing systemic circulation and inflammation. They may break the blood-brain barrier, stimulate microglia, which are the brain's immune cells, and cause neuroinflammation, all of which can result in neurotoxicity and neuronal injury. The buildup of oxidative stress, DNA damage, and dysregulated neurotransmitter systems brought on by exposure to air pollution can all lead to cognitive decline and neurodegeneration.

- ***Developmental Abnormalities and Reproductive Complications***: Chronic exposure to the environment can harm prenatal development and reproductive health. While chemical exposure from industrial byproducts and pesticides can affect the female as well as male reproductive well-being and fertility, early-life exposure to some environmental pollutants is believed to be linked to defects in development in babies. Toxins from the environment during pregnancy have been associated with a higher chance of infants developing abnormally. Endocrine-disrupting substances and heavy metals are a few samples of these contaminants.

During pregnancy, being exposed to heavy metals including lead and arsenic is linked to birth defects in offspring. These metals can pass through the placental barrier, altering the neurodevelopmental results, cognitive function, and behavior of the fetus. Endocrine-Disrupting Chemicals can prevent the endocrine system, which controls hormone synthesis and communication in the body, from operating normally. EDCs such as phthalates and certain pesticides have been associated with developmental anomalies in children, including negative effects on the body's immune system, neurodevelopment, especially the reproductive system.

Uncertainty exists regarding the precise methods via which environmental toxins affect development abnormally. These drugs, however, can impair vital embryonic development processes such as cellular proliferation, differentiation, and organ creation. They may cause oxidative stress, impair the operation of certain cellular receptors, interfere with hormone signaling, damage DNA and cause cellular malfunction (Balmes, 2010). The reproductive health and fertility of both men and women can be negatively impacted by chemical exposures from industrial pollution and pesticides. Reproductive issues have been linked to exposure to industrial pollutants such as polychlorinated biphenyls (PCBs), dioxins, and certain solvents. These pollutants have the potential to interfere with hormone signaling, impact sperm quality, and cause hormonal imbalances in women, which might result in infertility, menstruation problems, and unfavorable pregnancy outcomes. Problems with reproductive health in both men and women have been related to pesticide exposure, both occupational and environmental. Some pesticides can impact hormone levels, sperm quality, and normal reproductive processes. Exposure to pesticides has been linked to monthly cycle irregularities, infertility, and a higher chance of miscarriage in women (Zhong, 2016).

Environmental toxins and pesticides that are hazardous to reproduction do so through pathways that include hormone disruption, interference with gamete formation, oxidative stress, and DNA damage in reproductive organs. These outcomes may result in hormone imbalances, decreased sperm and egg quality, and problems with implantation and the growth of a healthy fetus.

6.2.2. Economic Impact

Due to higher healthcare expenditures, decreased productivity, and expenses associated with disabilities, these diseases can have a substantial negative impact on individuals, families, and society as a whole. Chronic environmental disorders frequently need continuing medical care, which can have a significant financial impact. People with these disorders may require ongoing medical care, prescription drugs, laboratory testing, imaging scans, and specialty therapies. The costs of managing chronic disorders can add up over time, especially if the illnesses call for long-term or permanent care. Chronic environmental illness sufferers may also experience more frequent hospitalizations and ER visits, which would increase healthcare expenses.

For people without sufficient health insurance or those who reside in areas with insufficient access to affordable healthcare services, the financial burden of healthcare bills can be particularly difficult. High out-of-pocket costs for drugs and treatments can put a strain on finances and might cause people to postpone or seek care insufficiently, which can affect both the health outcomes and financial well-being of individuals and families.

The capacity to work and be productive can be greatly impacted by chronic environmental conditions. The symptoms of various illnesses, such as respiratory issues, exhaustion, cognitive decline, and physical restrictions, can make it difficult for people to do their professions well. Reduced work hours, lost workdays, and even long-term incapacity may result from this.

Reduced productivity has effects on the economy at large in addition to affecting individuals' income. Economic productivity is lost when people with chronic environmental conditions are unable to work or are less productive. This may have an effect on the affected people as well as their employers, coworkers, and the workforce as a whole. Reduced worker engagement can slow social economic growth and result in decreased tax revenues for the government.

Chronic environmental illnesses may force sufferers to shell out more money for disability and functional restrictions. These costs could be incurred for the purchase or leasing of mobility aids like wheelchairs or walkers, as well as house adaptations. Individuals may also need support with everyday tasks or home healthcare services, which might (Balmes, 2010).

Disability-related expenditures can put a strain on people's finances, especially if their insurance doesn't cover them sufficiently or if public health programs don't pay them back. Additionally, accessibility upgrades and assistive technology can be costly, sometimes forcing people to make tough financial choices or look for financial aid to fulfill their demands (Zhong, 2016).

6.2.3. Quality of Life

Chronic environmental illnesses can have a significant negative effect on a person's quality of life. Chronic environmental illnesses frequently manifest with lingering symptoms that can have a big impact on everyday living. For instance, people with asthma may frequently suffer wheezing, coughing, shortness of breath, and chest tightness, which can restrict their ability to engage in physical activity and interfere with their sleep cycles. Environmental triggers like pollen, dust mites, or mold can cause allergic reactions that cause symptoms including sneezing, itching, runny noses, and skin rashes. These symptoms can be uncomfortable and interfere with daily activities. This illness, which may be influenced by environmental variables, can result in chronic tiredness, muscular discomfort, sleep disruptions, and cognitive challenges, which lower energy levels and restrict everyday activities.

Chronic environmental illnesses can make it difficult for people to participate in a variety of activities. For instance, individuals with chronic respiratory conditions may find it challenging to engage in strenuous physical activity or exert themselves as a result of breathlessness and diminished lung function. People who are sensitive to particular chemicals or environmental triggers might have to stay away from certain locations, such as extremely polluted areas or structures with strong chemical aromas, which might limit their ability to engage in social activities or travel without restriction.

Individuals may experience emotional anguish and problems with their mental health as a result of dealing with chronic environmental conditions. When you have a chronic environmental illness, you frequently worry and fret all the time about triggers and possible exacerbations. People may have increased alertness and dread due to ongoing worries about being exposed to allergies or toxins in the environment. Feelings of discomfort, restlessness, and an ongoing sensation of alertness might be caused by this ongoing state of worry.

Chronic environmental disorders can result in emotions of melancholy, hopelessness, and a loss of interest in formerly enjoyable activities due to their enduring symptoms, restrictions, and interruptions. People with the illness may feel confined and overburdened since it is persistent, which can lower mood generally and raise the chance of becoming uncomfortable.

Being emotionally taxed and distressed by the demands of a chronic environmental situation is common. People could struggle to adopt new habits, avoid triggers, and adhere to treatment regimens. Stress levels may rise as a result of the continuing need for environmental awareness and self-care, which might be harmful to mental health.

Some chronic environmental conditions require their sufferers to limit their exposure to specific environments or abstain from certain habits. People could avoid social gatherings, activities, or connections to protect their health, which could result in social isolation. The resulting social exclusion and loneliness can negatively affect mental health (Balmes, 2010).

People's sense of self and self-worth may be affected by chronic environmental conditions. People may, , feel that their sickness defines them or that their differences from others make them unique. As a result, there may be a sense of loss, a drop in self-esteem, and difficulties sustaining a positive self-image. Frustration and resentment can occasionally result from coping with the restrictions and difficulties brought on by chronic environmental conditions. People may become irritated by their inability to fully engage in activities, by others' lack of empathy or support, or by the unpredictable nature of their symptoms. These feelings might make people more stressed out and have a bad impact on their mental health.

Additionally, affecting cognitive function and causing problems with memory, focus, and attention are chronic environmental disorders. These cognitive difficulties may also exacerbate emotions of annoyance, tension, and a general decline in mental health. Chronic environmental conditions can also have an impact on social and interpersonal relationships. Certain chronic environmental disorders are poorly recognized, such as Multiple Chemical Sensitivity, which results in stigmatization and misinformation. This may make it difficult to connect socially and get the help you need. Relationships with family, friends, and coworkers may be impacted by the restrictions imposed by chronic environmental conditions. People may need to refuse invites to particular activities or request modifications at work (Zhong, 2016).

6.3 CAUSES OF CHRONIC ENVIRONMENTAL DISEASES

6.3.1. Air pollution

Numerous environmental variables and exposures have been linked to chronic environmental illnesses. The development and escalation of respiratory and cardiovascular disorders can be attributed to exposure to air pollutants such as fine particulate matter, nitrogen dioxide, oxygen, and volatile organic compounds.

Airborne particles with a diameter of 2.5 micrometers or less are referred to as fine particles. They can reach far into the respiratory system because they are so little. Asthma, bronchitis, and chronic obstructive pulmonary disease may be triggered or made worse by inhaling PM2.5 since it can irritate and inflame the respiratory system. Additionally, PM2.5 can enter the circulation and raise the risk of cardiovascular mortality and disorders including heart attacks and strokes.

Fine Particulate Matter is a byproduct of combustion, which includes fossil fuel heating in homes, power plants, industrial operations, and car emissions. It also contains secondary particles produced as a result of atmospheric chemical processes.

One of the main ingredients of photochemical smog is the gas ozone, which is made up of three oxygen atoms. Ozone exposure at high levels can irritate and inflame the respiratory system, resulting in respiratory symptoms such as coughing, sore throats, and difficulty in breath. Existing respiratory problems like COPD and asthma might get worse as a result.

Ozone has the potential to cause cardiovascular problems and raise the risk of heart attacks. Although it is not immediately released, ozone is created when volatile organic compounds as well as nitrogen oxides combine in the presence of sunlight.

Nitrogen dioxide is a reddish-brown gas that is predominantly produced during combustion processes and is frequently present in densely populated metropolitan areas. Nitrogen dioxide can aggravate the respiratory system and aggravate airway irritation. Long-term exposure to high concentrations of nitrogen dioxide can aggravate asthma symptoms, raise the risk of respiratory infections, and accelerate the onset of respiratory illnesses. Additionally, it has cardiovascular side effects including a higher chance of heart attacks and strokes. Vehicle emissions, power plants, industrial combustion processes, and household burning of fossil fuels are the main sources of nitrogen dioxide.

VOCs are a broad category of organic compounds that readily evaporate at room temperature and are present in a wide range of goods and activities. VOCs can irritate the respiratory system, the eyes, the throat, and the brain as well as produce nausea, migraines, and dizziness. Some VOCs, like formaldehyde and benzene, are proven carcinogens and can raise the chance of developing cancer. Additionally, VOCs might cause the onset or worsening of respiratory and cardiovascular conditions (Balmes, 2010).

VOCs can be released from a variety of sources, such as automobile emissions, industrial operations, solvents, paints, cleaning supplies, personal care items, and indoor air pollution from furniture and building materials.

6.3.2. Water Contamination

Water sources that are contaminated offer serious health hazards and can spread a number of diseases that are transmitted by water, including gastrointestinal infections and vector-borne illnesses. Pathogenic microorganisms including bacteria, viruses, and parasites can live in contaminated water and cause gastrointestinal diseases when consumed. *Escherichia coli* strains with a history of causing severe gastroenteritis have been linked to hemolytic uremic syndrome, hemorrhagic diarrhea, and abdominal discomfort.

An ailment caused by the *Salmonella bacterium*, salmonellosis is primarily spread by consuming tainted food or drink. *Salmonella* may be spread to people through the water when it is tainted, which can lead to illness. Animals, especially chickens, cattle, and reptiles, frequently have salmonella in their intestines. Water pollution may happen in a number of ways, including through inappropriate storage or distribution practices, sewage contamination, or runoff from agricultural operations.

When produce is washed or irrigated, or when food preparation surfaces, utensils, or equipment are contaminated, contaminated water can enter the human food chain. *Salmonellosis* can spread when people drink water infected with *Salmonella* or eat food that has been cooked with contaminated water. *Salmonellosis* normally takes 12 to 72 hours to incubate following exposure, however, it can occasionally take longer. *Salmonellosis* is most frequently characterized by diarrhea, frequently watery or bloody, stomach pains, fever, and vomiting. There may also be other symptoms including fatigue, headaches, painful muscles, and nausea. The severity and duration of symptoms, which can range from mild to severe and normally last 4 to 7 days, vary depending on the individual (Zhong, N. S. 2016).

Salmonellosis can occasionally result in more serious problems, such as bloodstream infections or the spread of the bacteria to other body locations, especially in susceptible populations including newborns, the elderly, or those with compromised immune systems.

An infectious condition known as campylobacteriosis is caused by the *Campylobacter bacteria*, particularly *Campylobacter jejuni* and *Campylobacter coli*. In the entire world, it is one of the most prevalent bacterial causes of gastroenteritis. Human infection can occur as a result of *Campylobacter* contamination of water. Animals, especially chickens, cattle, and other domesticated animals, frequently have campylobacter bacteria in their intestines.

Water pollution may happen in many ways, including through inappropriate storage or distribution practices, sewage contamination, or runoff from agricultural operations. Infection can result from drinking water infected with Campylobacter or eating food cooked with contaminated water (Balmes, 2010).

Campylobacteriosis normally takes 2 to 5 days to incubate following exposure, however, it can occasionally take longer. The most typical signs of campylobacteriosis are diarrhea, which can occasionally be bloody or watery. The lower abdomen is frequently the source of abdominal discomfort or cramping, which is also prevalent. Another typical symptom that often ranges from mild to moderate is fever. Other symptoms that may be present include tiredness, headaches, nausea, and other digestive issues. Symptoms might vary in intensity and length from person to person, but they often continue for a week.

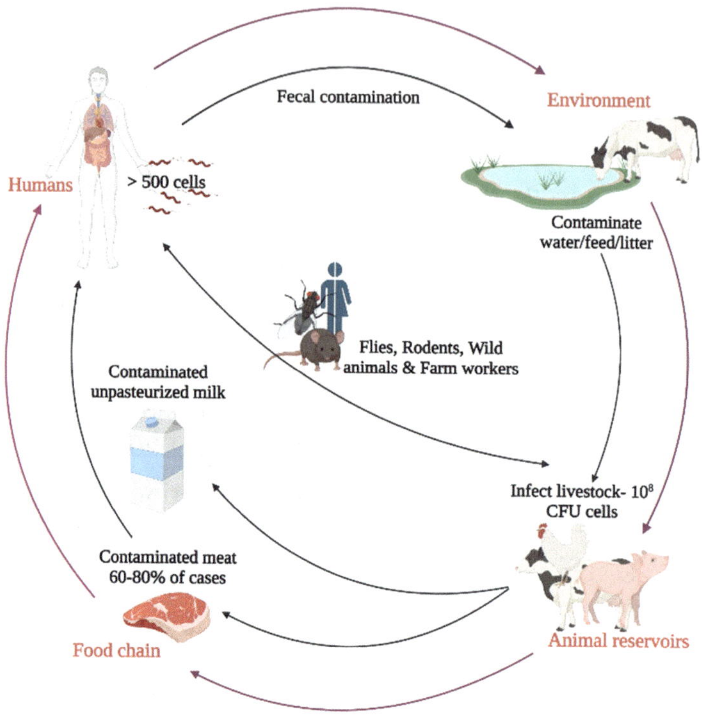

Figure 6.5: *Campylobacter* cycle.

Source: I Al Hakeem WG, Fathima S, Shanmugasundaram R, Selvaraj RK. *Campylobacter jejuni* in Poultry: Pathogenesis and Control Strategies. *Microorganisms*. 2022; 10(11):2134. https://doi.org/10.3390/microorganisms10112134

The disease known as cryptosporidiosis is caused by the tiny parasite *Cryptosporidium*. It is renowned for its capacity to live in a variety of water sources and is one of the main causes of waterborne illness globally. Protozoa, which are single-celled parasites, include Cryptosporidium. It may infect a broad variety of animals, including people, and has a complicated life cycle. There are various types of *Cryptosporidium*, with *Cryptosporidium parvum* harming people the most frequently.

Ingestion of infected food or drink is the main way that *Cryptosporidium* is spread. Indirect contact with infected people or animals can potentially spread the parasite. The parasite's latent and resistant form, Cryptosporidium oocysts, may persist in water for extended periods, making it very robust and challenging to eradicate with standard water treatment techniques.

Common signs of cryptosporidiosis are digestive problems. Watery diarrhea, which can be copious and protracted, is the most typical symptom.

Other signs and symptoms include nausea, vomiting, dehydration, exhaustion, and weight loss. The infection can be severe and life-threatening in those with weaker immune systems, such as those with HIV/AIDS or those receiving chemotherapy (Balmes,. 2010).

Giardia is a tiny parasite that can lead to giardiasis, an infection of the intestines. It frequently appears in water sources that have been tainted by human or animal waste. Giardia in water is frequently linked to subpar sewage systems, insufficient sanitation procedures, and feces-contaminated water sources. The protozoan parasite Giardia, also known as *Giardia lamblia*, is a member of the genus *Giardia duodenalis.* The active form, known as a trophozoite, and the dormant form, known as a cyst, both have a distinctive pear-shaped look. The cyst form is immune to abrasive environmental factors, such as chlorine disinfection, which is frequently employed in water treatment procedures (Zhong, 2016).

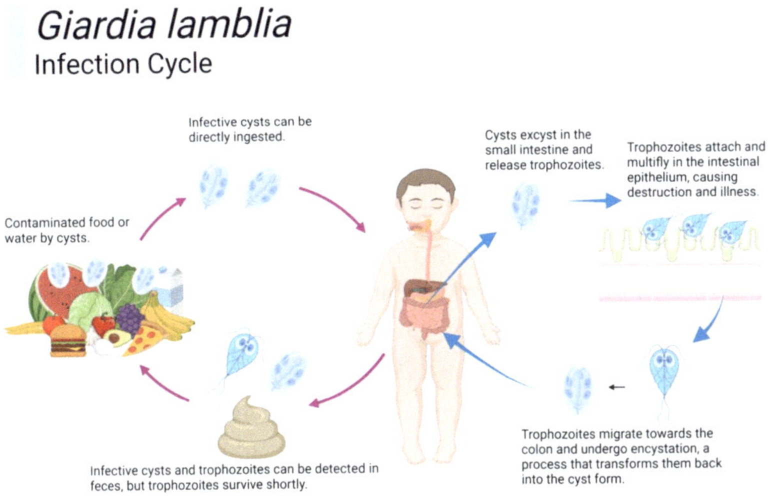

Figure 6.6: Giardiasis determinants.

Source: Khattak I, Yen W-L, Usman T, Nasreen N, Khan A, Ahmad S, Rehman G, Khan K, Said MB, Chen C-C. Individual and Community-Level Risk Factors for Giardiasis in Children under Five Years of Age in Pakistan: A Prospective Multi-Regional Study. *Children*. 2023; 10(6):1087. https://doi.org/10.3390/children10061087 [accessed 9 Jun, 2023]

Ingestion of tainted food or water is the main method of Giardia transmission. Direct touch between people or contact with infected surfaces can potentially spread the parasite. Giardia cysts can last for a very long time in water sources like rivers, lakes, or wells, posing a constant risk to the water's quality.

Symptoms of giardiasis usually appear 1 to 3 weeks after exposure. Diarrhea, which can be watery, oily, or foul-smelling, is the most typical symptom. Other signs and symptoms include nausea, exhaustion, bloating, gas, stomach cramps, and weight loss. Diagnosis and treatment might be difficult in certain circumstances since people may have sporadic diarrhea or no symptoms at all.

Water sources may act as hatching grounds for a variety of disease-transmitting vectors, including snails and mosquitoes. An important part of the Plasmodium parasite, which causes malaria, is transmitted by mosquitoes. Millions of people worldwide suffer from malaria, a potentially fatal illness that is most prevalent in tropical and subtropical areas.

The *Plasmodium parasite*, of which *Plasmodium falciparum* is the deadliest species, is what causes malaria. The parasite is mostly spread to people through the bites of female Anopheles mosquitoes that are infected. The infective stage of the parasite calledsporozoites, is injected into the circulation when an infected mosquito bites a human.

Once inside the human body, the *Plasmodium parasite* invades liver cells, where it multiplies and develops into mature forms known as merozoites. These merozoites are released into the bloodstream and invade red blood cells, where they replicate further. This cycle of invasion, replication, and release leads to the characteristic symptoms of malaria (Balmes, 2010).

Malaria symptoms normally develop 10 to 15 days following a mosquito bite; However, they can take longer to manifest. Recurrent fever, chills, sweating, headache, discomfort in the muscles and joints, exhaustion, nausea, and vomiting are some of the most typical symptoms. Depending on the Plasmodium species involved and the person's immunity, the severity of symptoms might vary. Malaria can have serious side effects, including organ failure, severe anemia, cerebral malaria, and even death. Malaria infections often exhibit periodic fever patterns due to the synchronous release of merozoites from infected red blood cells. Malaria symptoms normally develop 10 to 15 days following a mosquito bite; however, they can take longer to manifest. Recurrent fever, chills, sweating, headache, discomfort in the muscles and joints, exhaustion, nausea, and vomiting are some of the most typical symptoms. Depending on the Plasmodium species involved and the person's immunity, the severity of symptoms might vary. Malaria can have serious side effects, including organ failure, severe anemia, and cerebral malaria. The fever patterns of various Plasmodium species vary, with tertian fever for *Plasmodium vivax* and *Plasmodium ovale*, and quartan fever for *Plasmodium malariae*. The most hazardous type of Plasmodium, *Plasmodium falciparum*, can produce irregular or persistent fever patterns (Zhong, 2016).

The dengue virus is spread by *Aedes* mosquitoes, notably *Aedes aegypti* and *Aedes albopictus*. The viral disease dengue fever is common in tropical and subtropical areas. When an infected person is bitten by an *Aedes* mosquito, the mosquito gets the dengue virus. Following bites from infected mosquitoes can infect healthy people with the virus because it multiplies inside the insect. The dengue virus is a member of DENV-4. the four different serotypes of the *Flaviviridae* family and Each serotype has the potential to cause dengue fever, and recurrent infections with various serotypes may raise the chance of the disease developing severely.

Freshwater snails of a few different species act as intermediary hosts for parasitic worms called schistosomes. Schistosomiasis is a condition marked by symptoms including fever, chills, coughing, stomach discomfort, and blood in the urine or feces. It can be contracted by coming into contact with water contaminated with these snails.

Stagnant water features, such as ponds, lakes, or reservoirs, are good breeding grounds for vectors. The growth of vectors and the spread of these illnesses are facilitated by poor water management, insufficient drainage systems, and a lack of vector control methods. Chemical contaminants that are harmful to human health when ingested can also be found in contaminated water. Arsenic pollution of water can occur naturally or as a result of industrial activity in some areas. Long-term exposure to arsenic-contaminated water is linked to many health issues, including cancer, heart and respiratory issues, skin sores, and skin lesions. Water supplies close to mining or industrial sectors may contain harmful heavy metals including lead, mercury, and cadmium. Chronic exposure to these heavy metals via tainted water can result in organ damage, difficulties with development, neurological abnormalities, and some types of cancer.

6.3.3. Chemical Exposures

Chemical exposures can cause serious health problems, both in work environments and while using consumer items. Exposure to dangerous substances over an extended period or repeatedly can cause many chronic illnesses. Chemical handling is common in many sectors, including industry, agriculture, construction, and healthcare. Workers in these fields may come into touch with, breathe in, or consume harmful chemicals. Chronic illnesses may develop based on the specific chemical, exposure dose, and length of exposure. Examples of chemicals connected to occupational exposures include asbestos, which has been linked to cancer and lung ailments, benzene, which has been linked to leukemia, lead, which has been linked to developmental and neurological problems, and certain pesticides, which have been linked to a variety of health consequences (Zhong, 2016).

The usage of common consumer goods including cleaning supplies, cosmetics, and home furnishings can potentially expose one to chemicals. These goods contain chemicals including phthalates, parabens, formaldehyde, and volatile organic compounds. Chronic health problems can be exacerbated by prolonged or recurrent exposure to certain substances, whether by inhalation or direct touch. For instance, certain phthalates and parabens have been linked to hormonal disruption and complications during pregnancy, while prolonged exposure to VOCs might have negative effects on the nervous system and the lungs (Balmes, 2010).

Numerous forms of cancer have been linked to occupational exposure to chemicals such asbestos, benzene, formaldehyde, chromium compounds, and certain pesticides. Cancers of the lung, bladder, liver, kidney, and skin are only a few examples of them. Long-term or recurrent exposure to carcinogens can harm DNA and impair biological functions, which can result in the growth of malignant cells.

Neurological problems can occur from chemical exposure's harmful effects on the nervous system. Heavy metals like lead, mercury, and manganese, for instance, can build up in the body and result in neurotoxicity. Cognitive deficits, developmental delays, behavioral modifications, and even neurodegenerative diseases like Parkinson's disease might result from this. The central nervous system can be impacted by solvents like toluene and trichloroethylene, which are typically encountered in industrial environments.

The endocrine system, which controls the synthesis and action of hormones, is known to be disturbed by many substances. Both males and females may experience reproductive problems as a result of exposure to certain endocrine-disrupting substances. For example, certain phthalates and bisphenol A (BPA) have been connected to hormonal abnormalities, decreased fertility, menstruation problems, and unfavorable pregnancy outcomes. EDC exposure in males may have an impact on sperm quality and quantity, impacting fertility.

Certain dangerous compounds that are inhaled can harm the respiratory system. Asbestos, silica, coal dust, and other airborne pollutants are examples of compounds that can cause respiratory conditions including asthma, chronic obstructive pulmonary disease, and occupational lung illnesses when exposed at work. These ailments may result in ongoing respiratory discomfort, diminished lung capacity, and long-term lung damage.

New research indicates that certain chemicals may play a role in the development of cardiovascular disorders. Heart attacks, strokes, and other cardiovascular diseases have been linked to air pollution, particularly particulate matter and volatile organic compounds. Furthermore, some occupational exposures, including those to specific solvents and heavy metals, have been connected to high blood pressure, heart function issues, and a higher risk of cardiovascular issues.

6.3.4. Occupational Hazards

Employees may be more likely to acquire chronic illnesses as a result of occupational exposure to specific drugs or settings. Workers might be exposed to asbestos fibers in fields including construction, shipbuilding, and insulation installation. Long-term exposure to asbestos fibers can cause illnesses like asbestosis. Due to its heat resistance and insulating qualities, asbestos, a naturally occurring mineral, was often employed in a variety of industries. Asbestos fibers may have been inhaled by workers in industries including construction, shipbuilding, mining, and insulation installation. These tiny fibers can accumulate in the lungs over time as a result of inhalation (Balmes, 2010).

When asbestos fibers are inhaled, the lung tissue may become inflamed and scarred. The sharp, long-lasting fibers cause persistent irritation and damage to the sensitive lung tissues. This ongoing irritation eventually sets off a fibrotic reaction that causes the lungs to develop scar tissue (fibrosis). Asbestosis causes the lung tissue to scar, which causes the lungs to lose some of their flexibility and elasticity. It becomes more challenging to breathe as a result of the gradual stiffness and decreasing lung function caused by this.

Breathlessness is a potential symptom of asbestosis, particularly when exercising. Even routine duties might become difficult as the illness worsens. A persistent, dry cough is frequently seen with asbestosis. The cough may linger and get worse over time, making it uncomfortable and challenging to clear the airways. Certain asbestosis sufferers may feel as though their chests are tight or heavy. Due to the rigidity of the lung tissues and the decreased lung capacity, this is the case. Asbestosis compromises the respiratory system's ability to clear mucus and foreign particles effectively. The scarred lung tissue provides an environment conducive to the growth of bacteria, making individuals with asbestosis more susceptible to respiratory infections. These infections can exacerbate breathing difficulties and further impair lung function (Zhong, 2016).

Asbestosis has a long latency period, often taking several decades for symptoms to manifest after the initial asbestos exposure. The severity and progression of the disease can vary depending on the duration and intensity of exposure. While asbestosis is not reversible, early detection and prevention of further exposure can help slow its progression and manage symptoms effectively.

Asbestosis also increases the risk of developing other asbestos-related diseases, including mesothelioma (a cancer of the lining of the lungs or other organs), lung cancer, and pleural thickening (thickening of the lining around the lungs). These diseases often have a poor prognosis and are associated with significant health challenges.

Mesothelioma is a type of cancer that primarily affects the lining of the lungs (pleura), abdomen (peritoneum), or other organs such as the heart (pericardium) or testes. It is considered an aggressive and rare cancer that is closely linked to exposure to asbestos fibers. When asbestos fibers are inhaled or ingested, they can penetrate the protective linings of various organs, causing chronic inflammation and genetic damage to the cells. Over time, these damaged cells can develop into tumors. Mesothelioma specifically affects the mesothelial cells, which form the protective linings of organs (Balmes, 2010).

Lung cancer is a malignant tumor that starts in the cells of the lungs. Occupational asbestos exposure is a well-established cause of lung cancer, and when combined with smoking, it significantly increases the risk. Asbestos fibers are known to be carcinogenic, meaning they have the potential to cause cancer. The fibers can cause DNA damage and lead to the development of malignant cells in the lungs. Lung cancer associated with asbestos exposure is often referred to as asbestos-related lung cancer or lung cancer caused by asbestos. The risk of developing lung cancer from asbestos exposure is significantly amplified when combined with smoking tobacco. The combination of asbestos exposure and smoking has a synergistic effect, meaning that the risk is higher than the sum of the risks associated with each factor individually. Smokers exposed to asbestos have a much higher risk of developing lung cancer compared to nonsmokers or smokers who were not exposed to asbestos (Zhong, 2016).

Occupational exposure to certain heavy metals and ionizing radiation poses significant health risks to workers in various industries. Heavy metals such as lead, mercury, cadmium, and arsenic, as well as chronic exposure to ionizing radiation, can result in severe health consequences. Lead is a toxic heavy metal that can cause lead poisoning when workers are exposed to it over an extended period. Occupational exposure to lead commonly occurs in industries such as battery manufacturing, mining, and metalworking. Lead poisoning affects multiple body systems, with the central nervous system being particularly vulnerable. Chronic exposure to lead can result in neurological damage, cognitive impairments, anemia, and kidney problems. The neurotoxic effects of lead can lead to developmental issues in children and cognitive impairments in adults.

Mercury, another toxic heavy metal, is commonly found in industries such as mining, dentistry, and chemical manufacturing. Chronic mercury exposure can have detrimental effects on the nervous system and various organs. Neurological and behavioral abnormalities are common in individuals with mercury poisoning, along with kidney damage and adverse effects on the cardiovascular system. Pregnant women exposed to mercury may also transfer it to their unborn children, leading to developmental issues.

Cadmium is present in industries such as battery manufacturing, metal plating, and welding. Prolonged exposure to cadmium can result in chronic kidney disease, respiratory disorders, and an increased risk of lung cancer. The kidneys are particularly susceptible to cadmium toxicity, leading to impaired kidney function and potentially irreversible damage (Balmes, 2010).

Occupational exposure to arsenic, commonly found in industries such as mining, smelting, and pesticide manufacturing, can lead to various diseases. Skin lesions, peripheral neuropathy, and an increased risk of lung cancer are among the health consequences associated with chronic arsenic exposure. Long-term exposure to arsenic has also been linked to an elevated risk of other cancers, including bladder, liver, and skin cancer. Workers in nuclear power plants, healthcare facilities, and radiography departments may be exposed to ionizing radiation. Chronic exposure to ionizing radiation carries significant health risks.

Prolonged exposure to ionizing radiation increases the risk of developing various cancers. Workers exposed to radiation may face an elevated risk of leukemia, thyroid cancer, lung cancer, and certain solid tumors. The damage to DNA caused by radiation can lead to mutations and the uncontrolled growth of abnormal cells, resulting in the development of cancer.

Repeated exposure to radiation can cause radiation dermatitis, a condition characterized by skin damage. Inflammation, ulcers, and delayed wound healing are common symptoms. Radiation dermatitis can significantly impact the quality of life for those affected, leading to discomfort and potential complications.

Chronic radiation exposure can also affect the genetic material. It increases the risk of genetic disorders in offspring, posing potential risks for future generations. The mutagenic effects of radiation on DNA can lead to hereditary diseases and abnormalities in the exposed individuals' progeny (Zhong, 2016).

6.3.5. Climate Change

Climate change is a complex and multifaceted phenomenon that affects various aspects of the environment, ecosystems, and human health. One significant impact of climate change is its influence on chronic environmental diseases. Climate change is associated with an increase in the frequency and intensity of extreme weather events, such as hurricanes, floods, droughts, and wildfires. These events can have profound effects on human health and well-being. They can disrupt essential infrastructure, including healthcare systems, and lead to displacement, injuries, and loss of life. In the aftermath of extreme weather events, individuals may face challenges in accessing healthcare, medications, and necessary resources, which can worsen chronic conditions (Balmes, 2010).

Rising global temperatures are causing more frequent and severe heat waves. Heatwaves can have a substantial impact on individuals with chronic diseases, particularly cardiovascular and respiratory conditions. Heat stress can exacerbate existing conditions, increase the risk of complications, and even lead to life-threatening events such as heatstroke. Vulnerable populations, including the elderly, children, and those with pre-existing health issues, are particularly at risk.

Climate change can alter the geographical distribution and prevalence of certain diseases. Vector-borne diseases, such as malaria, dengue fever, Lyme disease, and Zika virus, are particularly influenced by climate factors. Changes in temperature, precipitation patterns, and the availability of suitable habitats for disease-carrying vectors can impact disease transmission dynamics. As a result, individuals with chronic conditions may face an increased risk of exposure to these diseases and their associated complications.

Climate change contributes to the deterioration of air quality, primarily through the emission of greenhouse gases and the formation of ground-level ozone. Poor air quality can have detrimental effects on respiratory health, particularly for individuals with chronic respiratory conditions such as asthma, chronic obstructive pulmonary disease (COPD), and allergies. Increased air pollution levels can trigger respiratory symptoms, exacerbate existing conditions, and lead to an increased risk of respiratory infections.

Climate change affects food and water security, leading to changes in the availability, quality, and accessibility of these essential resources. In regions experiencing droughts, floods, or changes in precipitation patterns, access to safe and nutritious food and clean water can be compromised. Inadequate nutrition and waterborne diseases can further exacerbate chronic conditions, weaken immune systems, and hinder overall health.

Climate change and its associated impacts can also have profound effects on mental health. The stress, trauma, and displacement caused by extreme weather events, loss of livelihoods, and disruptions to communities can contribute to anxiety, depression, post-traumatic stress disorder (PTSD), and other mental health disorders. Individuals with chronic conditions may experience heightened psychological distress, which can impact their overall well-being and ability to manage their conditions effectively (Zhong, 2016).

6.4 RESPONSES TO CHRONIC ENVIRONMENTAL DISEASES

6.4.1. Prevention and Awareness

Strategies for prevention, management, and treatment are used in response to chronic environmental illnesses to lessen their impact. Public health initiatives aiming at limiting exposure to environmental dangers and encouraging healthy habits have prevention and awareness as their main foundations. These initiatives concentrate on informing local populations, putting regulations into place, and pushing for adjustments that can reduce environmental risks and enhance all-around wellbeing. The following are some crucial specifics outlining the significance and methods used in preventative and awareness initiatives (Balmes, 2010). Public health organizations and agencies aim to increase community knowledge of any potential environmental concerns. This involves educating people about pollution sources such as vehicular exhaust, indoor pollutants, industrial emissions, and tainted water sources. By raising awareness, people will be better able to comprehend the possible health effects of these exposures and take the necessary precautions to lower their risks.

Communities must be made aware of the unique environmental dangers they may face and the preventive steps they might take as part of prevention initiatives. This includes sharing knowledge about the negative impacts of pollution exposure on health, offering advice on how to reduce exposure, and supporting healthy behaviors. To provide widespread access to information, educational campaigns can be carried out through a variety of channels, including public meetings, workshops, school programs, community events, and internet resources.

Strategies for prevention place a strong emphasis on the necessity of implementing safe habits to reduce exposure to environmental risks. This might involve suggestions for lowering indoor air pollution, such as good ventilation and refraining from smoking inside. Additionally, it can entail advocating safe food handling procedures to prevent foodborne diseases, encouraging workers to utilize protective equipment in high-risk occupational environments, and educating people on the proper handling and storage of chemicals. Communities may lessen their exposure to dangerous chemicals and related health concerns by supporting these activities.

Public health initiatives urge people to change their lifestyles in a way that lessens their susceptibility to environmental risks. This includes encouraging positive lifestyle choices including frequent exercise, eating a balanced diet, and abstaining from tobacco and excessive alcohol use. These lifestyle changes can improve immune function, promote general health, and lower the risk of chronic illnesses that may be made worse by environmental exposures (Zhong, 2016).

Promoting and executing laws that address environmental dangers and safeguard the public's health are part of prevention initiatives. This entails creating guidelines and standards for hazardous waste management, workplace safety, and other areas that influence environmental health. Strong regulations may assure the execution of preventative measures, restrict exposure to dangerous chemicals, and enforce pollution controls. To ensure a safe and healthy environment, public health organizations also endeavor to monitor compliance with rules and impose fines for noncompliance. When communities actively participate in decision-making processes and have a voice in environmental health issues, prevention initiatives are most successful. Public health organizations seek to include communities in debates, get their opinions, and provide them with the tools they need to take action. This might entail collaborating with neighborhood groups, running outreach activities, and including locals in the planning and execution of environmental health projects. Engaging communities encourage a sense of ownership and group accountability, resulting in more effective and long-lasting preventative measures.

6.4.2. Environmental Regulations

Governments and regulatory agencies establish environmental regulations as policies and rules to safeguard sustainable behaviors and protect the environment. These rules are intended to prevent environmental risks from occurring, protect natural resources, and promote human health (Balmes, 2010).

To reduce the amount of pollutants released into the atmosphere, governments set air quality guidelines. The allowable amounts of pollutants including sulfur dioxide, nitrogen oxides, particulate matter, and volatile organic compounds are specified by these regulations. To comply with these rules, businesses and power plants frequently have to install pollution control equipment and follow strict emission restrictions.

Water bodies including rivers, lakes, and coastal regions are intended to be protected from contamination by water quality standards. They outline the permitted concentrations of different pollutants in water, such as pathogens, pesticides, heavy metals, and industrial chemicals. To assure the quality of discharged water, businesses and wastewater treatment facilities must adhere to these criteria and conduct correct treatment procedures.

To guarantee the proper handling, storage, and disposal of hazardous substances, governments put rules into place. Guidelines for labeling, packing, shipping, and emergency response are all covered by these standards. They also outline specifications for things like suitable ventilation, containment techniques, and strategies for spill response for chemical storage facilities.

The correct handling and disposal of hazardous items are governed by hazardous waste rules. Specific processes for waste characterization, storage, transportation, and treatment are required for industries that produce hazardous waste. To keep an eye on the production, processing, and disposal of hazardous waste, governments frequently impose licenses and mandate regular reporting (Zhong, 2016).

The purpose of occupational health and safety regulations is to protect workers from environmental threats while they are at work. These laws set out standards for controlling exposure to hazardous materials, noise levels, and ergonomic risks. Additionally, they mandate that companies maintain safe working conditions, provide safety training programs, and supply personal protective equipment.

Large-scale development projects must comply with the regulation requiring environmental impact assessments. Before approval, they entail evaluating a project's possible environmental effects. The evaluation takes into account elements including water and air pollution, habitat degradation, and the depletion of natural resources. It aids in the identification of mitigating strategies to reduce harmful environmental consequences and advance sustainable development.

Regulations are in place in many nations to safeguard and preserve biodiversity. These rules may call for the creation of protected areas, the development of conservation strategies for threatened species, and limitations on the trade in endangered animals and plants. To reduce the negative effects on ecosystems, governments also support sustainable practices in sectors like forestry, agriculture, and fisheries.

To encourage the use of clean and sustainable energy sources, governments may establish renewable energy standards. These criteria frequently demand that a specific volume of power come from renewable resources like wind, solar, hydro, or geothermal energy. They may encourage the creation and deployment of technology for renewable energy, which would reduce greenhouse gas emissions and reliance on fossil fuels (Balmes, 2010).

6.4.3. Research and Surveillance

Understanding the connections between environmental exposures and chronic illnesses requires extensive research and observation. These actions offer insightful information on the frequency, patterns, and effects of environmental risks on human health. Establishing causal links between environmental exposures and chronic illnesses requires research. To learn more about how certain environmental dangers, affect human health, scientists perform epidemiological studies, lab tests, and observational studies. These investigations aid in establishing dose-response correlations, identifying the processes through which exposures result in the development of illness, and figuring out the cumulative effects of long-term exposures (Zhong, 2016).

Policymakers and public health professionals may create targeted initiatives to reduce the hazards associated with environmental exposures by recognizing these causal links. The particular health effects of environmental exposures are revealed by research. This involves researching the links between exposure to toxins, pollutants, and other environmental variables and the onset or worsening of chronic illnesses. For instance, scientists look at how exposure to heavy metals affects neurological problems, how air pollution affects respiratory illnesses, or how pesticide exposure affects cancer risk. Public health experts can determine the scope of health consequences, pinpoint susceptible populations, and create successful preventative and intervention measures by doing thorough research.

Systems of surveillance are essential for locating high-risk groups and regions where chronic environmental illnesses are more common. The systematic gathering, analysis, and interpretation of health data, such as illness incidence, prevalence, and death rates, are all part of surveillance. Public health organizations may see patterns, identify new trends, and focus actions on groups that are at risk by monitoring this data. Inequalities in disease burden among other demographic groups, such as racial or ethnic minorities, low-income areas, or occupational cohorts exposed to particular dangers, are also identified through surveillance (Balmes, J. R. 2010).

Continuous monitoring of illness patterns linked to environmental exposures is made possible by surveillance systems. Researchers and public health professionals can assess the efficacy of initiatives and policies targeted at lowering environmental hazards by monitoring changes in illness incidence or prevalence over time. The creation of evidence-based policies is aided by surveillance data, which also highlights gaps in interventions and gives feedback on the effectiveness of preventative initiatives. Additionally, monitoring enables prompt reaction and control actions by enabling early detection of disease outbreaks or emergent health hazards associated with environmental exposures.

Findings from research and monitoring are extremely important for guiding decision-making and developing policy. To create laws, set exposure limits, and provide recommendations for environmental protection, policymakers rely on scientific data. The creation of environmental laws, occupational safety standards, and public health programs can all be influenced by research findings. When selecting initiatives, allocating resources, and assessing the effectiveness of legislation intended to reduce environmental hazards, surveillance data are an invaluable resource. Environmental health is a subject where innovation and knowledge growth are driven by ongoing research and surveillance initiatives. Research discoveries add to the body of knowledge in the scientific community, allowing for the collection of data and promoting cooperation between academics and practitioners. This information helps the selection of novel therapies to prevent and treat chronic environmental illnesses, makes it easier to create new methods for assessing exposure, and informs current research projects.

6.4.4. Recommendation: Health Services and Support

To address and manage chronic environmental illnesses, health services and support are essential. The accurate diagnosis, treatment, and continuous management of these disorders are facilitated by an efficient healthcare system, readily available healthcare services, specialist clinics, and support networks. The infrastructure and knowledge required for the precise diagnosis and screening of chronic environmental illnesses are provided by healthcare services. Medical experts, such as doctors, nurses, and specialists, are essential in assessing patients' symptoms, carrying out diagnostic procedures, and deciphering test findings. They determine the connection between environmental exposures and the emergence of chronic illnesses using their knowledge and expertise. For those who may have been exposed to environmental risks, routine screenings and medical exams are especially crucial to help diagnose illnesses at the earliest stages, when treatment is most successful. Patients with chronic environmental disorders obtain the proper care and management techniques thanks to readily available healthcare services. Healthcare professionals collaborate closely with patients to create individualized treatment regimens that take the patient's ailment and severity into consideration. To reduce symptoms and enhance the quality of life, this may involve the administration of medications, treatments, rehabilitation, and lifestyle changes. To treat certain chronic environmental conditions, specialized clinics or departments may be set up, enabling all-encompassing and specialized care (Balmes, 2010).

Multiple healthcare providers must work together in a multidisciplinary manner to manage chronic environmental disorders. Pulmonologists, oncologists, neurologists, occupational health specialists, and other specialists who may offer specialized therapy and advice based on the particular disease and its associated environmental exposures may fall under this category. Collaboration across several healthcare specialties ensures a thorough and holistic approach.

Health care services provide families and patients with chronic environmental conditions information and counseling. Educating people about the disease, its causes, risk factors, and preventative strategies is a component of this. Healthcare professionals provide advice on how to treat symptoms, deal with the psychological and emotional effects of the disease, and develop healthy habits to reduce additional exposure to environmental dangers. Individuals are empowered by education and counseling to actively take part in their healthcare, make knowledgeable decisions, and follow treatment regimens (Zhong, 2016).

Resources and support groups are essential for helping people with chronic environmental illnesses and their families. People can interact, exchange experiences, and seek emotional support through community organizations, internet forums, and patient support groups. These networks also provide access to materials, information, and campaigns for environmental health activism. Support groups provide those afflicted by chronic environmental and diseases a sense of community, lessen feelings of loneliness, and encourage self-advocacy (Balmes, 2010).

In attempts to prevent and promote health about chronic environmental illnesses, health services are essential. Healthcare professionals provide advice on taking preventative actions, such as limiting exposure to environmental dangers, embracing a healthy lifestyle, and adhering to safety procedures in high-risk employment. By taking part in research, monitoring, and policy formulation efforts targeted at lowering environmental dangers and enhancing population health, they also support public health programs.

Chapter

7 LAND DEGRADATION AND ENVIRONMENTAL CHANGE

CONTENTS

7.1 INTRODUCTION

Land degradation is the process by which land loses its ability to support life due to human activity. This can happen through the introduction of invasive species, soil compaction, or over-grazing. It can also be caused by erosion from wind, water, or other factors. Environmental change is a broad term used to describe changes in the environment that cause problems for humans and other living things. Examples include rising sea levels or altered weather patterns resulting in drought or flooding. The process of land degradation and environmental change includes the loss of soil fertility, biodiversity and other natural resources as well as the loss of ecosystem services such as water supply, flood control and recreation. Land degradation is often caused by unsustainable farming practices or over-exploitation of natural resources such as forests, peatlands and marine ecosystems (Treidel, 2013).

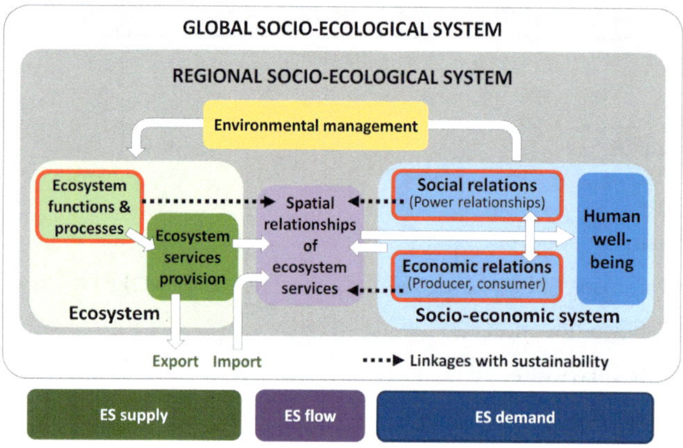

Figure 7.1: The current opinion on environmental sustainability.

Source: Schirpke U, Egarter Vigl L, Tasser E, Tappeiner U. Analyzing Spatial Congruencies and Mismatches between Supply, Demand and Flow of Ecosystem Services and Sustainable Development. *Sustainability*. 2019; 11(8):2227. https://doi.org/10.3390/su11082227 [accessed 12 Jun, 2023]

Global warming is a major cause of land degradation. When plants and animals can no longer survive in an area due to climate change, it becomes unable to support them. In addition to causing environmental damage, land degradation can also affect people's health by causing respiratory problems or increasing the risk of disease transmission. Land degradation is also bad for businesses because it can lead to decreased productivity and less revenue from farming, which can cause even more land degradation (Herold, M. 2017).

7.2 CAUSES OF LAND DEGRADATION AND ENVIRONMENTAL CHANGE

Land degradation is a term that describes the process of environmental change on the land. It can be caused by many factors, including climate change and human activity. Causes of land degradation are often categorized as natural or anthropogenic. Natural causes include weather patterns such as flooding, erosion, and subsidence. Anthropogenic causes include deforestation and agricultural practices that cause soil compaction. A contributing factor to land degradation is not always clear-cut; some factors may contribute to degradation while others may lead to conservation. For example, deforestation can lead to habitat loss for wildlife species that live in forests and savannas but at the same time, it can also lead to increased carbon storage in soils and forests (Treidel, 2013).

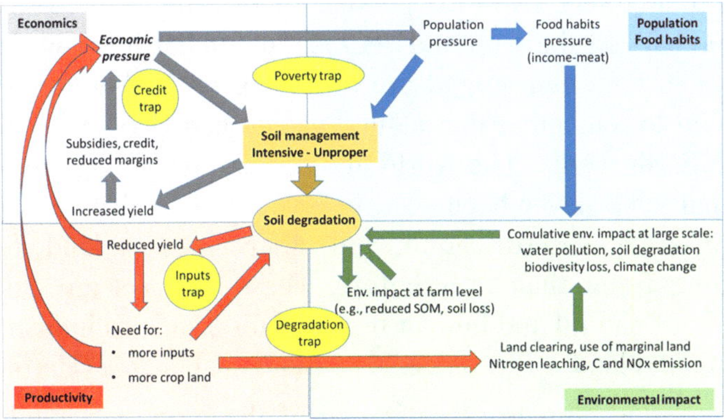

Figure 7.2: Impact of land degradation on the environment.

Source: Gomiero T. Soil Degradation, Land Scarcity and Food Security: Reviewing a Complex Challenge. *Sustainability*. 2016; 8(3):281. https://doi.org/10.3390/su8030281 [accessed 12 Jun, 2023]

Land degradation and environmental change are the results of many factors, but what we call a cause depends on its effect on the land. For example, if a river floods and kills all the trees in an area, then we would say that the flood was a cause. However, if there were no trees, to begin with, then the flood would not be considered a cause. The identification of causes began in the 1960s with scientists who were studying how climate change affected life on Earth. They found that certain factors—like deforestation—caused other factors such as drought and erosion (i.e., soil loss). Some factors are driven by human behavior, such as overgrazing or deforestation. These factors can affect the health of the soil and ultimately lead to land degradation. Other factors are natural disasters or other types of climate change that damage the soil by washing away topsoil or changing its composition. In some cases, a combination of both human and natural causes leads to a degradation in the quality of our environment. This can cause problems for humans and wildlife alike by altering habitats (for example, through changes in vegetation) or making it difficult for animals to survive in an area (for example, by changing river flow patterns).

Natural Causes

Natural causes refer to the changes in an ecosystem that are brought about by natural forces. Examples of natural causes include deforestation and erosion. When we talk about natural causes, we're talking about things that happen in nature without direct help from people. Since millions of years ago, these things have changed the land and ecosystems of Earth. Land degradation and changes in the environment are often caused by human actions, but it is important to remember that natural causes can also have a big impact on how the Earth works. The world and society are facing environmental problems that have never been seen before. Climate change, biodiversity loss, deforestation, desertification, and other types of land degradation are happening faster than ever before. These changes are caused by a complex mix of natural and human-made factors, such as industrialization, urbanization, and land use practices that aren't sustainable (Herold, 2017).

The worsening climate crisis is one of the most important things to worry about in the world right now. Rising global temperatures, which are mostly caused by people releasing greenhouse gases, are having a wide range of effects on ecosystems, such as a rise in sea level, more extreme weather, and changes in the way rain falls. These changes have big effects on farming, water supplies, biodiversity, and where people live. Loss of biodiversity is another important problem that needs attention. The loss of species and ecosystems can be caused by things like natural disasters and disease outbreaks. But things that people do, like destroying habitats, overusing resources, and polluting the environment, have sped up the rate of species going extinct. This has led to imbalances in ecosystems and the loss of valuable ecological services (Treidel,H. 2013).

Land degradation caused by things like soil erosion, volcanic eruptions, and geological processes is also a big problem. Even though these are natural processes, they can be made worse by human actions, which speeds up the damage. Unsustainable agricultural practices, deforestation, and bad land management can make ecosystems and communities more vulnerable to natural causes of land degradation and make their effects worse. Society as a whole is becoming more and more aware of how important it is to deal with these problems. On both the national and international levels, people are working to lessen the effects of land degradation and environmental change caused by natural and human factors. Initiatives include ways to manage land sustainably, conservation efforts, policy frameworks, and international agreements that aim to lower greenhouse gas emissions and help protect biodiversity.

- ***Geological Process***: One cause is geological processes that occur in nature including earthquakes, landslides, and erosion due to wind or water erosion caused by glaciers melting down. These include compaction (which occurs when heavy objects press down on topsoil), subsidence (when areas of earth drop into lower levels), subsidence/trenching (when an underground river or aquifer drains its water), and subsideral collapse (when an underground fault breaks). Volcanic eruptions are caused by magma moving toward the surface of the earth. The magma cools as it rises through the Earth's crust, causing it to solidify into rock. When this solidified lava reaches the surface, it erupts into a volcano. Explosions from an eruption can raise mountains or create craters in the ground. These are called calderas (Herold,. 2017).

  Earthquakes: Earthquakes are caused by movement within the earth's crust due to pressure from underneath or overlying rock layers or fluids such as groundwater flowing through cracks or faults in rocks (see diagram below). Earthquakes can cause landslides and tsunamis if they strike near coastlines; they can also cause subsidence (collapse) of landforms like volcanoes and faulted strata formations such as coal seams which may cause flooding downstream from affected areas (Treidel, 2013).
- ***Climate Change***: Climate change refers to long-term changes in how the weather works and how warm or cold it is on Earth. It is mostly caused by things that people do, like burning fossil fuels, cutting down trees, and making things in factories that release greenhouse gases into the air. These greenhouse gases, such as carbon dioxide (CO_2), methane (CH_4), and nitrous oxide (N_2O), trap heat from the sun and cause the Earth's temperature to rise.

This has many natural, social, and economic effects. The rise in temperatures around the world is one of the most important effects of climate change. Over the past 100 years, the average surface temperature of the Earth has risen by about 1 degree Celsius, with large differences between areas. Because of this trend, the polar ice caps and glaciers are melting, which is making the sea level rise. Because of this, low-lying coastal areas and small island countries are at risk of flooding, which could force people to move and cause many species to lose their homes.

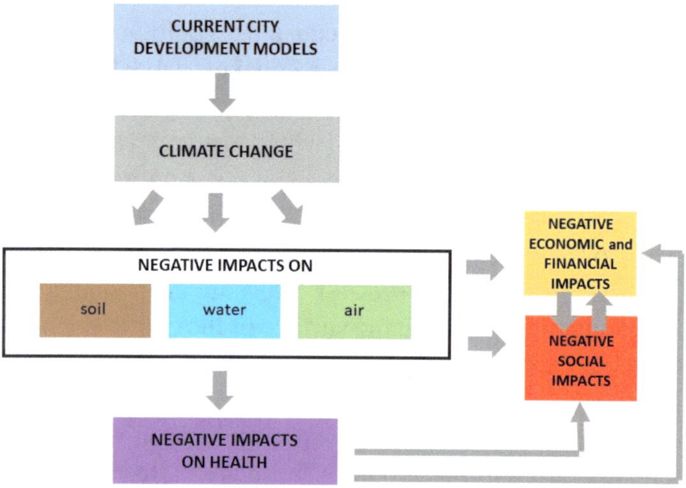

Figure 7.3: Climate change and its impacts.

Source: Girard LF, Nocca F. Climate Change and Health Impacts in Urban Areas: Towards Hybrid Evaluation Tools for New Governance. *Atmosphere*. 2020; 11(12):1344. https://doi.org/10.3390/atmos11121344

Changes in weather trends and a rise in extreme weather events are also big effects of climate change. Heat waves, floods, and wildfires are getting worse and happening more often, which is bad for people's health, crops, and ecosystems. On the other hand, there are more heavy rainstorms and floods, which damage infrastructure, stop food production, and spread diseases that are carried by water. Climate change has effects on more than just the earth. It also has social and economic effects. Changes in temperature and rainfall trends can throw off agricultural systems, which can lead to less food being grown and less food being available. This can make food insecurity worse and make communities more vulnerable, especially in poor countries that depend a lot on agriculture. The health of people is also at risk because of climate change. As the regional range of mosquitoes that carry diseases like malaria and dengue fever expands, rising temperatures can make the spread of vector-borne diseases like malaria and dengue fever worse. Extreme heat events can also make it more likely for people to get sick from the heat or die from it. This is especially true for the old and people who already have health problems (Treidel, 2013).

Climate change has a big effect on the economy. Costs can be high for things like building structures that can withstand climate change, switching to renewable energy sources, and putting in place sustainable land management practices. On the other hand, the economic benefits of doing something about climate change, like creating green jobs, promoting renewable energy businesses, and making energy use more efficient, offer chances for economic growth and long-term development. To deal with climate change, people all over the world need to work together. As stated in the Paris Agreement, the international community has agreed that it is very important to cut down on greenhouse gas pollution and keep global warming to well below 2 degrees Celsius above pre-industrial levels. This agreement aims to make the world's reaction to climate change stronger by reducing greenhouse gas emissions, adapting to the effects of climate change, and helping developing countries with money and technology.

- ***Weather Extremes***: Extreme weather, like strong storms, heat waves, droughts, and floods, is a major cause of land degradation and changes in the environment. Climate change has made these extreme weather events happen more often and with more force. This has bad effects on both natural processes and human activities. Hurricanes, typhoons, and cyclones are all examples of strong storms that can do a lot of damage to the land and the environment. Strong winds can knock down trees, kill plants, and erode the topsoil, which can lead to soil erosion and the loss of rich land. Heavy rain from these storms can cause flooding, which makes land erosion and sediment buildup in rivers and lakes worse. Pollutants and trash can be carried by floodwaters, which can pollute water sources and and damage aquatic environments (Herold, 2017).

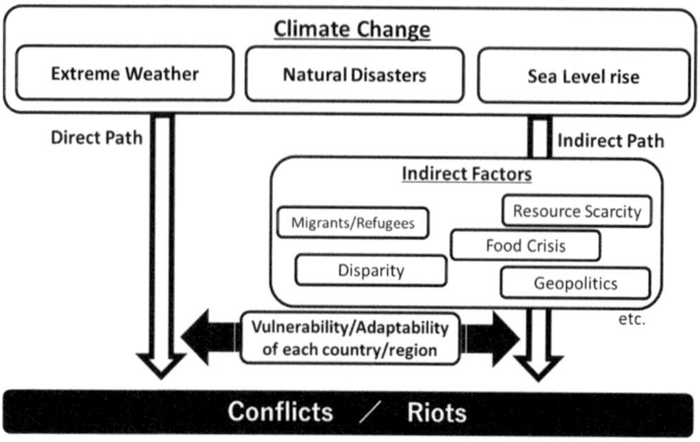

Figure 7.4: Observed and projected impacts from extreme weather events.

Source: Sekiyama T. Climate Security and Its Implications for East Asia. *Climate*. 2022; 10(7):104. https://doi.org/10.3390/cli10070104

Heat waves, which are times of high temperatures that last for a long time, can hurt the land and plants. Too much heat can cause grounds to dry out and lose water, which can lead to drought conditions. Droughts make it harder for plants and farmers to get water, which can kill crops, lower output, and damage the land. In the worst cases, floods can lead to desertification, which is when once-fertile land becomes dry and like a desert. On the other hand, heavy rain and storms can also damage the land. Heavy rain can wash away soil and nutrients and change the environment. This runoff can cause the topsoil to wash away, which is important for plant growth and keeping the soil fertile. Floods can leave sediments on farmland, which can bury crops and make the soil less useful. A sudden flood can also mess up natural environments and habitats, forcing animals to move and hurting biodiversity (Treidel, 2013). Extreme weather has direct and indirect effects on human actions and ways of making a living. For crop output, agriculture relies a lot on stable weather patterns. Extreme events can disrupt farming practices, which can lead to lower yields and economic losses. In coastal places, erosion, and coastal erosion can be caused by strong storms and floods. This puts infrastructure, homes, and communities at risk. The loss of fertile land, lack of water, and damage to ecosystems can also affect food stability, water availability, and human health (Herold, 2017).

Anthropogenic Causes

Anthropogenic causes are the effects and actions of people that lead to land destruction and changes in the environment. Most of these problems are caused by what people do, such as industry, urbanization, deforestation, farming methods, pollution, and taking natural resources. Human-made factors are now the most important ones that shape the Earth's ecosystems, and they have had big and far-reaching effects on the environment. Landscapes and natural environments have changed a lot because of industrialization and urbanization. As industries and towns grow, natural land is often turned into built-up areas. This leads to the loss of habitat, the splitting up of ecosystems, and the destruction of ecosystems. Construction work, building infrastructure, and setting up transportation networks all add to the degradation of land, the compacting of soil, and the disruption of natural water systems.

Deforestation, which is mostly caused by expanding farms and cutting down trees, has big effects on land degradation and environmental change. When trees are cut down on a large scale, the soil erodes, biodiversity is lost, and local and regional climate patterns change. When trees are cut down, forests are less able to take in carbon dioxide, which contributes to greenhouse gas emissions and global warming.

Land degradation has also been caused by farming methods, especially those that are too heavy and can't be kept up indefinitely. Chemical fertilizers, pesticides, and herbicides can pollute the soil and water when they are used. This can lead to soil degradation, less fertile soil, and polluted water. Soil erosion, desertification, and the loss of agricultural output are all made worse by bad land management practices like overgrazing, improper irrigation, and growing only one crop.

Pollution damages land and the environment from a variety of sources, including industrial emissions, poor waste management, and the discharge of hazardous chemicals and pollutants. . Pollutants in the air can fall to the ground and contaminate the soil and plants. Industrial waste water, farming runoff, and improper waste disposal can all pollute water bodies and put ecosystems and human health at risk. Natural resource extraction and use, like mining, oil drilling, and gas research, have big effects on the environment.

These things often destroy habitats, cause land erosion, pollute water, and cause greenhouse gases to be released into the air. Also, taking resources out of an area can upset local groups, cause fights over land rights, and have long-term social and economic effects (Treidel, H. 2013).

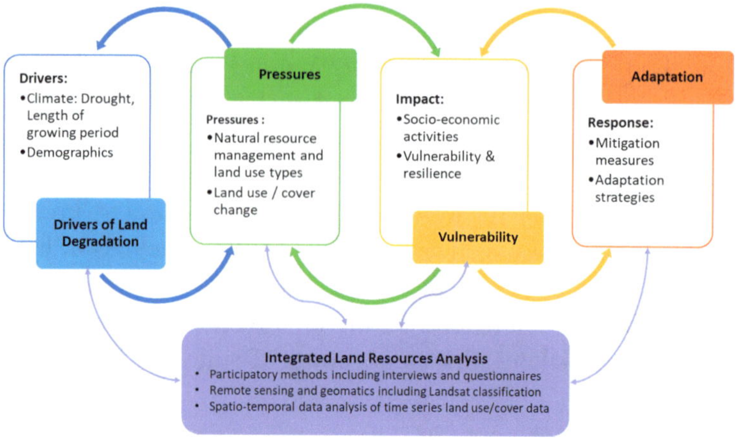

Figure 7.5: Land degradation.

Source: Gobin A, Hien LTT, Hai LT, Linh PH, Thang NN, Vinh PQ. Adaptation to Land Degradation in Southeast Vietnam. *Land*. 2020; 9(9):302. https://doi.org/10.3390/land9090302 [accessed 12 Jun, 2023]

Human-made causes of land decline and environmental change have led to a loss of biodiversity, more greenhouse gas emissions, climate change, and many other problems for ecosystems and human well-being. It is important to understand and deal with these causes to reduce their effects and encourage practices that are good for the environment. This means putting in place rules and policies that put conservation, healthy land management, reforestation, reducing waste, and switching to renewable energy sources at the top of the list. Raising knowledge, promoting education, and getting people to act in ways that are better for the environment are also important ways to deal with anthropogenic causes and create a more sustainable future (Herold, 2017).

Human activities such as logging, quarrying, and agricultural practices lead to land degradation because they destroy forests. When forests are destroyed, they can no longer keep water contained in their root systems. This causes soil erosion and leads to flooding. In addition, when trees are cut down for logging purposes or fuel wood, they release carbon into the atmosphere and contribute to global warming. However, not all human activities lead to land degradation. Some practices can improve the quality of soil and reduce erosion by creating new plants that help hold the soil together. For example, when farmers plant seeds that grow into plants they help prevent erosion by keeping the soil from blowing away from its roots or letting water run through them instead of through channels carved into the ground's surface (called ditches or furrows). In addition to human activity and natural processes such as wildfires or hurricanes that damage forests through windstorms or flooding (which cause landslides), there are also natural causes of land degradation such as climate change which has led to changes in weather patterns that affect agriculture production levels (Treidel, 2013).

7.3 DESERTIFICATION

Desertification is a complicated process that includes the degradation of land in arid, semi-arid, and dry sub-humid areas, turning once-fertile land into conditions similar to a desert. It is mostly caused by a mix of nature and human-made things. Climate changes, like floods and changes in the way it rains, are natural causes, while overgrazing, cutting down trees, and farming in the wrong way are human-made causes. Desertification has a wide range of effects on both the environment and people's lives. Desertification can lead to the loss of plant cover, soil erosion, less water availability, and the depletion of groundwater supplies in arid and semi-arid areas. This, in turn, can lead to less food security, less food production, and the relocation of people whose livelihoods depend on the land. Desertification also leads to the loss of unique plant and animal species because they can't live in the harsh desert conditions.

Several tactics and approaches have been made to stop and reverse desertification. Sustainable land management is an important part. This includes things like saving the earth, collecting water, planting trees, and spreading the word about agroforestry systems. The goal of these practices is to improve soil fertility, increase plant growth, and make more water available. This will help restore land that has been damaged and make ecosystems stronger. Implementing policies and rules that encourage the use and control of land sustainably is another important step. This means putting measures to avoid and reduce desertification into national development plans, making laws to stop land degradation, and encouraging local communities to help with land restoration. Desertification is a world problem that needs everyone to work together to solve it. Therefore, international cooperation and collaboration are also very important.

Also, it is important to make people aware of desertification and teach them about its causes and effects. This can give people the tools they need to manage their land sustainably, take part in reforestation projects, and take steps to save water supplies. Investing in research and new technologies is also important to find new ways to restore land and to track and measure how well current efforts to stop desertification are working. Desertification is a big problem for dry and semi-dry places all over the world. It is caused by both natural and human causes, and it has very bad effects on both the environment and people's lives. But desertification can be stopped and even turned around with the help of sustainable land management practices, policy interventions, community involvement, and technological advances. By fixing up degraded land, working to protect wildlife, and making sure that natural resources are used sustainably, we can make ecosystems stronger and improve the lives of people who live in places where deserts are spreading (Herold, 2017).

7.4 SOIL EROSION

Soil erosion is a natural process that happens when soil particles break off, get moved by things like wind, water, and gravity, and then get dumped somewhere else. Even though erosion is a natural process, people have done a lot to speed it up, which is bad for agricultural output, water quality, and the health of water bodies. Soil erosion starts when small pieces of soil break away from the top of the earth. This can happen when raindrops fall, when water flows when the wind blows, or when people do things like till the soil or build something. Once the soil particles are free, they are moved by the eroding forces, which are mostly water and wind. When rainwater runs over the land, it picks up soil particles and carries them away. This is called water erosion. Wind erosion, on the other hand, happens when soil particles move because of the force of the wind, especially in dry, desert places (Treidel, 2013).

Both natural and man-made things can cause land erosion. Some natural factors are how much rain falls, how steep the slope is, the type of soil, and how much vegetation is there. When it rains hard and for a long time, especially on steep hills, the rate of erosion can go up. The type of soil, how it is put together, and how much organic matter it has all affect how likely it is to get washed away. Plants are very important for stopping erosion because they cover the ground and have roots that hold the dirt in place. Erosion rates are made worse by things that people do, like cutting down trees, not managing land properly, overgrazing, and using bad farming methods. When plants are taken away or land isn't taken care of well, the protected cover is gone, leaving the soil bare and open to erosion.

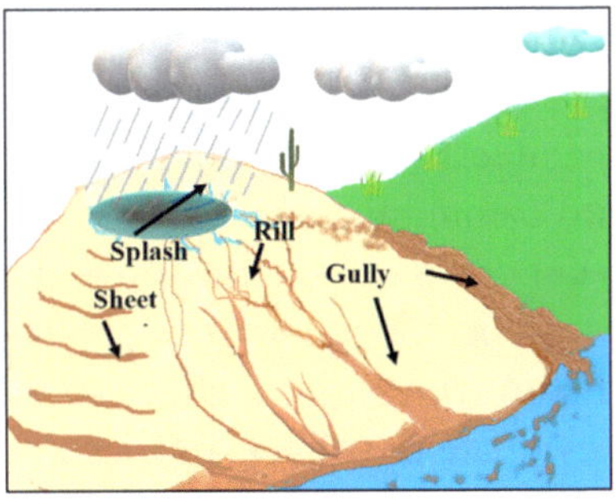

Figure 7.6: The various forms of soil erosion.

Source: Senanayake S, Pradhan B, Huete A, Brennan J. A Review on Assessing and Mapping Soil Erosion Hazard Using Geo-Informatics Technology for Farming System Management. *Remote Sensing*. 2020; 12(24):4063. https://doi.org/10.3390/rs12244063 [accessed 12 Jun, 2023].

Soil erosion has a lot of different effects on both the world and on the things people do. One of the biggest effects is on the amount of food that can be grown. When soils erode, they lose their healthy top layer, which has the nutrients plants need to grow. Because of this, crop rates go down, which makes farming less productive and makes food less secure. Erosion also changes the structure of the soil, making it less able to hold water and making droughts and storms more likely. This makes the problems farmers already have even worse and threatens the long-term health of farming systems. The quality of the water is also hurt by soil loss. As runoff water moves soil particles, they can end up in water bodies, which is called sedimentation. Sedimentation lowers the quality of water by making it cloudier, blocking sunshine, and taking away oxygen. It can also move soil particles with pollutants on them, like fertilizers and pesticides, which can get into water sources and hurt aquatic ecosystems. Also, erosion causes the loss of important topsoil, which takes thousands of years to form. This makes the soil resource even worse and makes it less able to help plants grow (Treidel, 2013).

To stop soil loss, you need to use more than one strategy. To stop erosion and keep the soil healthy, it is important to put soil conservation measures into place. On hills, techniques like contour plowing, terracing, and strip cropping slow down the flow of water and stop soil erosion. Conservation tillage methods, like no-till or limited tillage, disturb the soil as little as possible and keep the residue cover, which makes the soil more stable. Using the right land management techniques, such as rotational grazing and agroforestry systems, can also help stop runoff and improve the health of the soil. Also, to change people's behavior, It is important to make people aware of soil loss and its effects and to teach them about them. Encouraging farmers, landowners, and communities to use sustainable land management methods can do a lot to cut down on soil erosion and help protect the soil. It is also important for researchers, policymakers, and other interested parties to work together to come up with and use effective ways to stop erosion, as well as to monitor and evaluate the effects of these strategies on soil health and agricultural output.

7.5 DEFORESTATION

Deforestation is when forests or tree cover are taken out of a place on purpose or for good. It has become a major environmental problem with far-reaching effects on land degradation, the loss of forest habitats, carbon emissions, and climate change. Land pollution is one of the most obvious and direct effects of cutting down trees. Forests are very important for keeping the soil stable because they stop flooding and help water get into the ground. When trees are cut down, the protective canopy that protects the soil from heavy rain is taken away. This makes the soil more prone to erosion and makes it less productive. When the soil is open, it is more likely to be damaged by wind and water. This makes the land less good for farming and makes it unusable (Herold, 2017).

Deforestation also leads to the loss of forest ecosystems, which are very varied and do a lot of good for the environment. Forests are home to a huge number of plant and animal species, many of which are unique to the forest and are in danger of going extinct. Loss of species is directly caused by deforestation, which destroys their homes. This loss has big effects on how well the ecosystem works and how well it can recover since each species has a unique part to play in keeping ecological balance and helping the ecosystem as a whole stay healthy. Deforestation also releases carbon into the air and makes climate change worse. Through the process of photosynthesis, trees work as carbon sinks because they take in carbon dioxide from the air. When trees are cut down, the carbon they held is released back into the air as carbon dioxide. This adds to the amount of greenhouse gases in the air. Estimates show that deforestation is responsible for a big chunk of the world's carbon emissions. This makes it a major cause of climate change. Because of this, there are more greenhouse gases in the air, which causes temperatures to rise, changes in the weather, and bad effects on ecosystems and human groups. Deforestation also has indirect effects on climate change because it changes the weather trends in local and regional areas. Forests are very important for controlling weather and rainfall because they affect evaporation, cloud formation, and the amount of moisture in the air. When trees are cut down, these processes get messed up. This changes the way rain falls and makes people more likely to be affected by droughts, floods, and other extreme weather events. Changes in the climate have big effects on agriculture, water supplies, and human settlements, especially in areas that rely on forests. For the effects of deforestation to be fixed, we need a wide range of methods for conservation and sustainable forest management. Focus on getting people to use land responsibly, like cutting down on illegal logging, using sustainable logging methods, and promoting reforestation and afforestation programs. Protected areas and conservation parks are very important for keeping forest ecosystems and biodiversity alive, so they should be set up as soon as possible (Treidel, 2013).

Also, international agreements and frameworks, like the Reducing Emissions from Deforestation and Forest Degradation (REDD+) program, try to give developing countries incentives to cut down on deforestation and improve how forests are managed. These systems realize how important forests are for fighting climate change and give money to countries that take real steps to protect forests and cut carbon emissions. Education and awareness campaigns are also important for making people, communities, and businesses feel more responsible and encouraging them to use sustainable practices. Deforestation and its effects can be lessened by encouraging sustainable ways of living, supporting certified sustainable goods, and investing in green technologies (Herold, 2017).

7.6 LAND MANAGEMENT PRACTICES

Land management is a key part of preventing and reducing land degradation, which is important for keeping ecosystems healthy, boosting agricultural output, and making sure land use is sustainable. Sustainable land management practices include a wide range of methods and strategies that help people take care of and responsibly protect the land. This includes things like reforestation, agroforestry, conservation gardening, and terracing. Agroforestry is a way to use land that mixes growing crops or raising animals with planting and taking care of trees on purpose. Agroforestry is a method of farming that uses trees to make the land more productive. Trees help stop soil erosion by holding the soil in place with their roots. This makes wind and water erosion less likely to happen. They also help make the soil more fertile by recycling nutrients and adding organic matter. Trees also give crops and animals shade and protection, increase biodiversity, and store carbon dioxide, which helps slow climate change.

Conservation agriculture is a way of farming that encourages methods that are good for the soil and ecosystems and don't hurt them too much. It is based on three main ideas: disturbing the soil as little as possible, keeping the soil covered all the time, and rotating or changing crops. By doing less tilling, farms can stop the soil from washing away and keep the soil's structure, organic matter, and good organisms. Keeping the soil covered with crop leftovers or cover crops helps keep water from washing away the soil and makes it easier for water to soak into the ground. This lowers the risk of runoff and drought. Crop rotation or diversification improves the way nutrients are used, lowers the number of pests and diseases, and makes farms stronger overall.

Terracing is a way to manage land that includes building flat platforms on slopes to stop water from running off and soil from washing away. By building steps, water is caught and kept, so it can slowly soak into the soil and stop it from washing away. Terracing works especially well in areas with hills or mountains where slopes are prone to erosion. It helps keep the earth moist, helps crops grow, and keeps the topsoil from washing away. Also, erosion control and water saving can be improved by combining terraces with vegetative measures like grass strips or contour planting (Treidel, 2013).

Reforestation is the process of putting trees back where they had been cut down or killed. It means planting trees or letting them grow back on their own to bring back the forest cover and ecosystem services. Reforestation helps stop land degradation by stopping soil erosion, making it easier to keep water in the ground, and protecting wildlife. Trees help keep hills stable, reduce the effects of wind and water erosion, and protect ecosystems that are easily damaged. They also make it easier for ecosystems to do things like store carbon, filter water, and make places for animals to live (Herold, 2017).

These methods for sustainably managing land have more benefits than just keeping the land from getting worse. They help reduce the effects of climate change, protect biodiversity, handle water resources, and make farming more productive. To put these practices into action, you need a mix of policies that support them, skill building, technical assistance, and the participation of stakeholders. Through education, training programs, and financial help, it is important to make farmers and landowners aware of these practices and encourage them to use them.

7.7 LAND DEGRADATION AND FOOD SECURITY

Land degradation makes it hard to grow food and can have bad effects on crop output, farmers' incomes, and food security around the world. As the world's population keeps growing, it becomes more important to make sure food production is safe. Understanding how land degradation affects food security is important for coming up with good ways to deal with this problem. Crop output going down is one of the most important ways that land degradation hurts food production. Land degradation is caused by things like soil erosion, nutrient depletion, and salinization. These things make farming land less fertile and less productive. Soil runoff, for instance, gets rid of the top layer of soil, which is full of nutrients and organic matter that plants need to grow. This reduces the availability of nutrients, the soil's ability to hold water, and its structure, all of which can impact food yields. Nutrient depletion and salinization make these problems even worse, making it hard for crops to get the water and minerals they need to grow well.

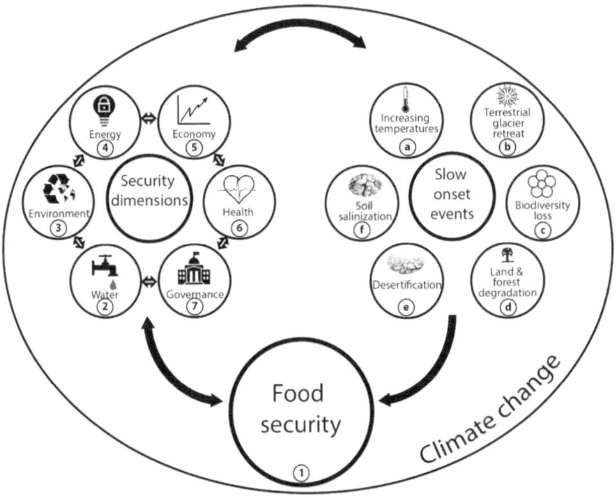

Figure 7.7: Land degradation and challenges of food security.

Source: G Stavi I, Paschalidou A, Kyriazopoulos AP, Halbac-Cotoara-Zamfir R, Siad SM, Suska-Malawska M, Savic D, Roque de Pinho J, Thalheimer L, Williams DS, et al. Multidimensional Food Security Nexus in Drylands under the Slow Onset Effects of Climate Change. *Land*. 2021; 10(12):1350. https://doi.org/10.3390/land10121350

When crop yields go down, it affects more than just one farmer. It affects whole agricultural processes and people's ways of making a living. Lower yields can mean less money and food for farms, which can make them poorer and less likely to have enough to eat. Small-scale farmers, who depend a lot on farming to make a living, are especially at risk from the effects of land degradation. Farmers can get stuck in a circle of poverty and needing help from outside sources if they lose farmland or get less food from their crops. Land degradation also has bigger effects on food security around the world. As agricultural output goes down, it becomes harder to meet the growing demand for food around the world. Land decline affects both how much food is grown and how well it is grown. Nutrient depletion in soils can lead to crops that don't have enough nutrients, which can affect the health of people who depend on these crop produce. Also, land degradation can mess up local and regional food systems, leading to unstable markets and price changes that can make food poverty even worse (Treidel, 2013).

Getting rid of the link between land degradation and food security will take more than one thing. To stop and reverse land degradation, the first thing that needs to be done is to use sustainable land management methods. Techniques like terracing and contour plowing, as well as agroforestry and the use of organic fertilizers, are used to protect the land. Farmers can keep or increase food yields by using these methods to improve soil fertility, save water, and reduce soil erosion. Another important thing is to make it easier for people to use tools and technologies. Small farmers don't always have access to things they need, like fertilizers, better seeds, and watering systems. Farmers can be much more productive and resistant to land degradation if they have access to these tools and can take part in training and capacity-building programs.

Food security can also be improved by encouraging farming systems that are both diverse and healthy. Diversifying crops, doing mixed farming, and using integrated pest management can make a farm more resistant to pests, diseases, and changes in the weather. These methods make it less likely that crops will fail and give people a more stable and varied food source. Research and development must be funded in order to come up with new and safe ways to farm. This includes making crop types that can handle climate changes, making it easier to use water efficiently, and spreading the word about precision agriculture. Also, if you want to stop land degradation and make sure there is enough food for everyone, you need effective laws and governance systems that put sustainable land management first and help small-scale farmers (Treidel, 2013).

7.8 POLICY AND GOVERNANCE FOR LAND MANAGEMENT

Policy and administration are very important for dealing with land degradation, promoting sustainable land use, and making sure that natural resources are protected. Effective policies, rules, and government systems set up a framework for managing land, balancing different interests, and protecting the environment. In the context of land degradation and environmental change, they are important for achieving sustainable development goals and reducing the effects of human activities on ecosystems. By setting standards and rules for how land can be used, one of the main jobs of policies and laws is to stop and control land degradation. There may be zoning rules, plans for how land is used, and environmental impact studies to make sure land is managed in a way that is sustainable and responsible. By making clear rules and requirements, policies can discourage harmful practices like cutting down trees, overgrazing, and changing the way land is used in a way that isn't good for the environment while pushing more sustainable ones.

Also, policies and rules can use economic and legal methods to encourage people to manage land sustainably. Farmers and landowners can be encouraged to use sustainable methods like agroforestry, conservation agriculture, and soil conservation by giving them money in the form of subsidies or tax breaks. Permits and licenses are examples of regulatory tools that can make sure environmental standards are met and stop people from doing things that hurt the land. Land management policies need to be put into place and enforced with the help of good government systems. This includes making agencies and groups in charge of overseeing land use planning, making sure rules are followed, and enforcing laws. Effective governance requires that everyone knows their jobs and responsibilities and that everyone works together. To do this, government agencies, local communities, indigenous groups, non-governmental organizations, and private sector players may need to work together (Herold, 2017).

Participatory methods of governance are also very important because they allow different points of view and local knowledge to be taken into account when making decisions. Engaging stakeholders in the formulation and application of land management policies help people feel like they own the land, improves transparency, and makes it more likely that the policies will work. Local groups, especially those whose lives depend directly on the land, should be able to help make decisions that affect their lives and make sure their needs are taken into account. Also, foreign cooperation and agreements are very important for fixing land degradation around the world. Countries that are facing similar problems can share information, best practices, and technical know-how by working together. International conventions like the United Nations Convention to Combat Desertification (UNCCD) and the Convention on Biological Diversity (CBD) provide a framework for global action and cooperation to fight land degradation and promote sustainable land management (Treidel, 2013).

7.9 LAND DEGRADATION AND WATER RESOURCES

Land degradation has a big effect on water supplies. It makes water scarce, pollutes water, and uses up freshwater sources. Land degradation, especially in the form of soil erosion, deforestation, and wrong ways to use land, has a direct effect on the amount and quality of water, which is a big problem for both people and wildlife. One of the most important ways that land degradation affects water supplies is by making less water available. Soil erosion can be caused by things like intense farming, cutting down trees, and building cities. This can lead to the loss of topsoil, which is important for keeping water in the ground and letting it soak in. As a result, more rainwater runs off, which means less water goes back into the ground and less water flows through rivers and streams. This can lead to a lack of water in the long run, especially in places where water comes from the ground.

Water pollution can also be caused by land degradation. When soil erosion happens, sediments and pollution from farming, like fertilizers and pesticides, are carried away by runoff and dumped in water bodies. This sedimentation and pollution of water sources impact the quality of water, making it unsafe for people to drink and bad for marine ecosystems. Too many nutrients from farming runoff can also cause eutrophication. This makes algae grow too much and uses up all the oxygen in water bodies, which hurts their ecological health even more. Land degradation also makes it harder to get enough fresh water.

Deforestation, for example, throws off the natural water cycle by decreasing evaporation and changing the way rain falls. Trees are very important because they take in water, help clouds form, and control how much rain falls. Without enough tree cover, the hydrological cycle is thrown off, which leads to less rain and water supplies drying up. Also, land degradation can cause wetland degradation and the destruction of natural water storage places, which makes it even harder to get fresh water (Herold, 2017).

To lessen the effect on water supplies, it is important to deal with land degradation. For water to stay available and of good quality, sustainable land management methods must be used. Afforestation and reforestation are two ways to bring back vegetation cover, make the land more stable, and help water get into the ground. Conservation farming methods, like using cover crops and doing less tilling, help keep water in the land and stop it from washing away. It is also important to improve how land is used and take steps to stop pollution. This can be done by building terraces, plowing along the edges of the land, or building check dams to slow down the flow and catch sediment. Using best practices in agriculture, like precision irrigation, proper fertilizer management, and agroforestry systems, can cut down on the amount of pollutants that get into water bodies, which improves the quality of the water (Treidel, 2013).

Also, in order to protect freshwater resources in the face of land degradation, it is important to have good strategies for managing water resources, such as water conservation, creating protected areas around water sources, and promoting integrated water management. For these methods to work, it is important to encourage local communities, farmers, policymakers, and water control authorities to participate and work together (Herold, 2017).

Chapter

8

CAUSES, EFFECTS, AND SOLUTIONS TO DESERTIFICATION

CONTENTS

8.1 INTRODUCTION

In arid, semi-arid, and dry sub-humid areas, desertification, a pervasive environmental problem, poses serious threats to ecosystems and human well-being. Research intends to dive into the factors that contribute to desertification, as well as its effects and potential remedies. A greater comprehension of this complex issue will be attained by examining scientific research and professional viewpoints.

Degradation of the land, the soil, and the vegetation that leads to the spread of desert-like conditions is known as desertification. The delicate balance between natural processes and human activity can easily be upset in arid places, where water availability is scarce and susceptible to weather changes. Desertification is a process that is accelerated by unsustainable land management practices, overuse of resources, climate change, and population pressures (Weltz, 2019).

Desertification is the process that causes a major reduction in production in arid, semi-arid, and dry sub-humid environments, resulting in the extension of desert-like conditions. Natural factors like droughts and climate change, as well as human-induced factors like deforestation, overgrazing, and poor farming practices, are the main causes of the degradation of land, soil, and vegetation. Beyond environmental issues, desertification affects socioeconomic stability, food security, water availability, and political dynamics.

A multimodal strategy including sustainable land management techniques, integrated water resource management, policy interventions, and international cooperation is needed to address the complex problems caused by desertification. Research intends to offer useful insights into preventing desertification and fostering sustainable development in vulnerable places by examining the origins, impacts, and potential remedies.

Such in-depth analysis of desertification will help communities, governments, and scholars better understand the problem and collaborate to create solutions that will lessen its consequences.

It is possible to restore damaged landscapes, protect ecosystems, and enhance the livelihoods of populations impacted by this environmental crisis by addressing the underlying causes of desertification and applying sustainable land and water management techniques (Runyan, 2013).

8.2 THE SCOPE AND IMPORTANCE

Over a billion people worldwide are affected by desertification, which affects about one-third of the Earth's geographical surface. In the impacted areas, it poses serious risks to ecosystems, biodiversity, food security, water resources, and socioeconomic stability.

The degradation of delicate ecosystems and loss of biodiversity are other significant environmental effects of desertification. The ability of ecosystems to offer crucial functions, such carbon sequestration, soil formation, and water control, is diminished as a result of the deterioration of the land and vegetation.

Beyond environmental issues, desertification has a significant negative impact. It poses a serious risk to the availability of food, especially in areas where agriculture is a major source of revenue and subsistence. Crop yields drop as productive land transforms into a desolate wasteland, and cattle grazing becomes unsustainable. This not only puts the livelihoods of farmers and pastoral communities in danger but also fuels migration, increases food insecurity, and poverty.

Water quality and water shortages are both exacerbated by desertification. As the landscape and vegetation deteriorate, soil erosion worsens, causing more material to wash into rivers, lakes, and reservoirs. This sedimentation impacts the availability of clean water for both human consumption and agricultural use and decreases water storage capacity. Groundwater resources become contaminated as a result of the loss of plant cover, which also lowers ecosystems' ability to naturally filter and absorb water.

The effects of desertification on society and the economy are significant. In areas severely afflicted by desertification, displaced populations, unstable economies, and social unrest are frequently seen. Poverty and inequality are made worse by the loss of fertile land and chances for employment, leading to social and economic inequities both within and between communities. Additionally, desertification can exacerbate already-existing conflicts over finite resources, escalating geopolitical tensions and hampering efforts to promote sustainable development (Weltz, 2019).

Given the size and importance of the problem, combatting desertification calls for an all-encompassing strategy that involves multiple parties. Sustainable land management techniques, integrated water resource management, policy interventions, and international cooperation should all be a part of effective initiatives. It is possible to improve the well-being of communities impacted by this urgent environmental crisis by restoring degraded landscapes, enhancing ecosystem resilience, and taking the necessary actions to address the core causes of desertification (Runyan, C2013).

8.3 DESERTIFICATION FACTORS

8.3.1 Global Warming and Climate Change

Arid regions are growing as a result of climate change's changing precipitation patterns and rising global temperatures. Increasing evaporation rates, protracted droughts, and shifting weather patterns make existing arid regions even more susceptible to desertification.

Elevated temperatures are a result of the phenomena of global warming, which is caused by the buildup of greenhouse gases in the atmosphere. Water scarcity in dry areas is exacerbated by higher temperatures, which speed up the evaporation of surface water and soil moisture. Due to vegetation's effort to live as water availability declines, the land becomes degraded and desert-like conditions start to invade.

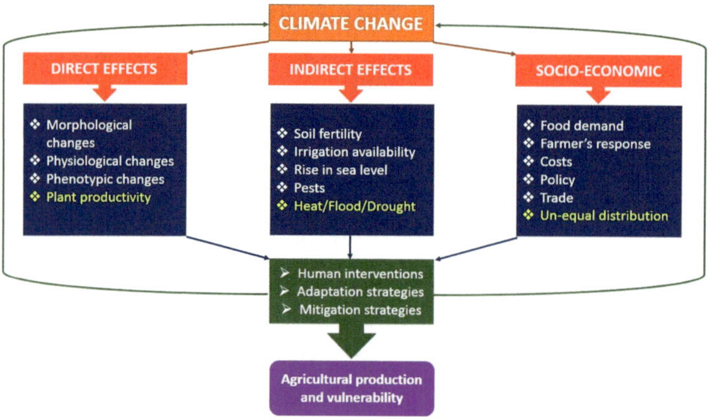

Figure 8.1: Global warming and crops.

Source: Raza A, Razzaq A, Mehmood SS, Zou X, Zhang X, Lv Y, Xu J. Impact of Climate Change on Crops Adaptation and Strategies to Tackle Its Outcome: A Review. *Plants*. 2019; 8(2):34. https://doi.org/10.3390/plants8020034 [accessed 8 Jun, 2023]

Changes in precipitation patterns, another effect of climate change, make desertification much more difficult. Reduced rainfall in some areas causes more frequent and severe droughts. Long-term water shortages impede agricultural endeavors, lower crop yields, and drive communities to move in pursuit of better job possibilities. In contrast, other places would have more intense rainfall, which would worsen soil erosion and topsoil loss, further reducing the productivity of the land.

A vicious loop exists between desertification and climate change. As the process of desertification advances, the loss of vegetation lowers ecosystems' ability to absorb carbon dioxide, a significant greenhouse gas that contributes to global warming. Thus, previously susceptible places will see worsening conditions as a result of climate change (Weltz, 2019).

Addressing the underlying causes of climate change and lowering greenhouse gas emissions are necessary to mitigate desertification. Arid regions can be less affected by climate change by switching to renewable energy, implementing energy-saving measures, and utilizing sustainable land management and agricultural methods. Effectively preventing desertification also requires increasing ecosystem resilience and taking steps to adapt to changing climatic circumstances.

Formulating successful policies and initiatives requires an understanding of the relationship between desertification and climate change. It is possible to lessen the susceptibility of dry regions to desertification and encourage sustainable development in these environmentally vulnerable areas by addressing climate change through mitigation and adaptation methods.

Overgrazing, especially when it involves cattle, exposes soil to erosion and removes vegetative cover. Desertification risk is increased by unsustainable farming practices such as inefficient irrigation techniques, excessive pesticide usage, and insufficient soil conservation measures (Runyan, 2013).

8.3.2 Land Clearing and DEFORESTATION

The ability of the land to retain moisture is decreased and the rate of soil erosion is accelerated by the clearance of forests and vegetation cover for logging, agriculture, and urban growth. Deforestation upsets the natural order, accelerating the aridification and desertification processes (Weltz, 2019).

Forests are essential for sustaining healthy soil and controlling water cycles. Raindrops are deflected by the dense tree canopy, which lessens their impact on the soil's surface and promotes slow water infiltration. Tree roots also compact the soil, limiting erosion and preserving its structure. These protective qualities are lost when forests are cut down, which increases soil erosion and water runoff. Additionally, removing land and deforestation damage ecosystems and contribute to the loss of species. A variety of plant and animal species can be found in forests, many of which have developed to fill particular ecological niches. These species experience population decreases and, in some cases, extinction as a result of habitat loss brought on by the clearing of forests. Ecosystem services, such as nutrient cycling and pest regulation, which are crucial for preserving soil fertility and agricultural output, are further disrupted by the loss of biodiversity (Runyan, 2013).

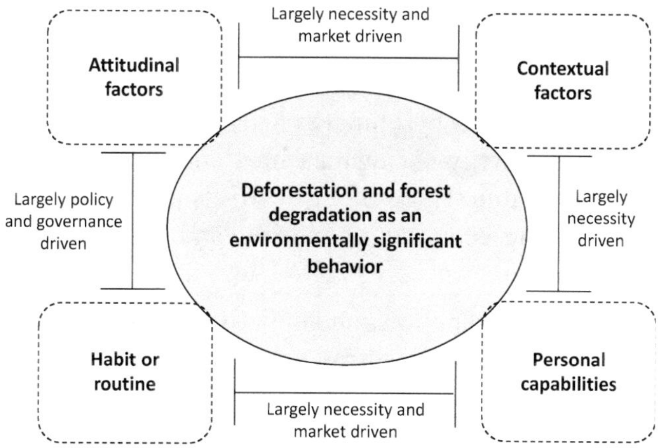

Figure 8.2: Conceptual framing of the aspects of deforestation and forest degradation (D&D) as an environmental behavior.

Source: Duguma, L., Atela, J., Minang, P., Ayana, A., Gizachew, B., Nzyoka, J., & Bernard, F. (2019). Deforestation and Forest Degradation as an Environmental Behavior: Unpacking Realities Shaping Community Actions. *Land,* 8(2), 26. MDPI AG. Retrieved from http://dx.doi.org/10.3390/land8020026

Furthermore, converting forests into agricultural land frequently entails irresponsible land management techniques like slash-and-burn farming. This method, which is frequently utilized in tropical areas, consists of clearing vegetation off the land and burning it to make room for cultivation. While it temporarily improves the soil's nitrogen content, exposure to the sun and wind that follows quickens soil evaporation and degradation. The need to further expand into forests arises as a result of the declining fertility of the land over time, which also affects agriculture yields.

Sustainable land management techniques must be used to reduce the effects of deforestation and land clearing on desertification. This includes initiatives to restore forest cover and encourage ecosystem regeneration, such as reforestation and afforestation. Furthermore, putting conservation techniques into practice can help protect vital habitats and maintain biodiversity. Examples of these efforts include protected areas and sustainable forest management.

Regulations against deforestation and other policy interventions, such as land-use planning, are also crucial. To lessen the strain on forests and address the causes of deforestation, sustainable agriculture practices, agroforestry systems, and support for alternative livelihood options can all be encouraged (Weltz, 2019).

It is feasible to stop the progression of desertification, rehabilitate degraded land, and support the long-term sustainability of ecosystems and livelihoods in impacted areas by addressing deforestation and land clearing practices.

8.3.3 Inefficient Use of Water and Drought

Water shortage and drought conditions in arid regions are caused by improper management of water resources, including ineffective irrigation systems, unrestricted groundwater exploitation, and insufficient water conservation measures. Prolonged droughts accelerate the process of desertification (Runyan, 2013).

Desertification is largely a result of poor water management. Inefficient irrigation techniques can lead to excessive water use and waste in dry locations where water is already scarce. Water loss and soil salinization can result from outdated irrigation techniques, such as flood irrigation, which have high evaporation rates and ineffective water distribution. In addition to reducing the amount of water resources available, this reduces soil fertility and increases the likelihood of erosion and degradation.

Desertification and water scarcity are also made worse by unrestricted groundwater extraction. Groundwater is frequently the main supply of fresh water in arid areas. However, excessive groundwater pumping that goes beyond what is sustainable causes aquifers to dry up, water tables to drop, and land to sag. The soil deteriorates and the beginning of desert-like conditions are caused by the shallow-rooted vegetation's inability to acquire enough water as the water table drops.

Desertification is also a result of poor water conservation efforts. Water shortage during dry spells is a result of inadequate water storage facilities and a lack of runoff management and rainwater harvesting equipment. The inability to properly save water increases how vulnerable ecosystems and communities are to droughts and desertification. In arid and semi-arid areas, droughts—characterized by protracted periods of low rainfall—occur naturally. But as drought events have become more frequent and intense, the effects on ecosystems and human livelihoods have gotten worse. Conditions of drought increase the risk of wildfires, hasten the loss of soil moisture, and limit vegetation cover. The ground becomes more vulnerable to erosion and degradation as a result, speeding up the desertification process.

Combating drought and poor water management is essential to halting desertification. The risk of desertification can be decreased by implementing effective irrigation systems, encouraging water conservation techniques, and enforcing rules on groundwater extraction. The long-term sustainability of arid regions depends on integrated water resource management, which includes equitable water distribution, sustainable use, and ecosystem-based techniques (Weltz, 2019).

8.3.4 Soil Degradation and Erosion

The topsoil is depleted of essential minerals and organic matter due to severe wind and water erosion and unsustainable land management methods. This erosion reduces the soil's fertility and capacity to hold onto moisture, hastening the development of desertification.

Desertification progresses largely as a result of soil erosion and deterioration. The soil is more susceptible to erosion by wind and water in dry and semi-arid areas where the natural vegetation cover is minimal. Fine soil particles can be carried away by strong winds, a process known as wind erosion, leaving behind an infertile and barren surface. Similar to water erosion, which occurs when rainwater washes away the topsoil and exposes the underlying soil layers, high rainfall, and poor soil conservation techniques can cause erosion (Runyan, 2013).

The natural equilibrium of the soil is disturbed by deforestation, excessive grazing, and improper farming techniques including intensive tilling and monoculture. By exposing the soil to the elements and removing the vegetative cover, the soil is less protected from erosive processes, which increases its susceptibility to erosion and degradation. Additionally, the depletion of nutrients and loss of organic matter further reduce the fertility of the soil, impeding plant development and hastening the processes of desertification.

The soil's capacity to hold onto moisture is reduced as the topsoil erodes. This causes a shortage of water and exacerbates the aridification processes by increasing surface runoff and reducing the infiltration of water. Additionally, the reduced agricultural output caused by the decreasing soil fertility makes it difficult for communities to maintain their standard of living in the impacted areas (Runyan, 2013).

Figure 8.3: Soil erosion and deterioration are exacerbated by unsustainable land management techniques.

Source: Wantzen KM, Mol JH. Soil Erosion from Agriculture and Mining: A Threat to Tropical Stream Ecosystems. *Agriculture*. 2013; 3(4):660-683. https://doi.org/10.3390/agriculture3040660

8.3.5 Urbanization and Human Population Growth

Urbanization and rapid population growth put a lot of strain on arid and semi-arid regions, expanding agricultural acreage, raising water needs, and depleting natural resources. These elements help weak ecosystems become degraded and turned into deserts.

Arid regions are significantly impacted by the rising population and urbanization trends. To meet the need for food as populations rise, agricultural production must be increased. This growth frequently encroaches on marginal territories, such as dry and semi-arid regions where there are already scarce soil and water resources. The delicate balance of these ecosystems is upset when natural areas are turned into agricultural fields, which increases their susceptibility to desertification (Weltz, 2019).

Urbanization increases the pressure on dry regions even more. Resources such as food, energy, and water are in greater demand as cities grow. Large amounts of water are frequently needed in urban areas for home and industrial uses. Urbanization's increased water demands could strain already-limited local water resources in arid locations, further harming ecosystems and hastening the process of desertification. In addition, infrastructure development includes the establishment of industrial facilities, buildings, and roadways. These actions frequently cause habitat fragmentation, biodiversity loss, and land degradation. Natural landscape disturbance increases the vulnerability of dry regions to desertification by fragmenting ecosystems and upsetting biological processes.

8.4 EFFECTS OF DESERTIFICATION

8.4.1 Effects on the Environment

- ***Reduction in Biodiversity***: A decrease in biodiversity is caused by desertification, which makes fewer plant and animal species' habitats suitable. Threats to arid-adapted species are growing, and the extinction of foundational species disturbs the integrity of ecosystems. The loss of biodiversity has an influence on ecosystem services like pollination, nitrogen cycling, and pest control in addition to the natural equilibrium of ecosystems.
- ***Desert Encroachment and Soil Degradation***: The soil becomes less fertile as desertification advances, making it harder for plants to grow. Desertification is exacerbated by the removal of vegetation cover, which leaves the soil vulnerable to erosion. By being transported away by wind and water, the eroded soil causes sediment to accumulate in nearby waterways and desert regions to encroach on adjoining landscapes. This process not only destroys the damaged land but also has a severe effect on the agricultural land and nearby ecosystems (Runyan, 2013).
- ***Altered Hydrological Systems***: Desertification reduces water infiltration and percolation into the soil by interfering with the normal hydrological cycle. Due to this modification, groundwater recharge, river flows, and surface runoff are all reduced, increasing water scarcity and the loss of aquatic habitats. Affected downstream regions are also impacted by altered hydrological systems, which reduces the amount of water available to ecosystems and human populations.
- ***More Frequent Sandstorms and Dust Storms***: As desertification reduces the amount of vegetation, the exposed soil is more vulnerable to wind erosion. This mechanism aids in the development of sandstorms and dust storms, which can have

significant effects on air quality, human health, and the long-distance transport of sediment. Dust storms can harm agriculture, infrastructure, and transportation systems in addition to reducing visibility and posing respiratory health problems.

Desertification has major, far-reaching effects on the ecosystem. Ecosystems, human health, and socioeconomic stability are all significantly impacted by biodiversity loss, soil degradation changed hydrological systems, and an increase in dust and sandstorms. To reduce the negative effects of desertification and improve the sustainable management of arid areas, it is imperative to address these environmental implications.

8.4.2 Social and Economic Effects

- ***Hunger and Undernourishment***: Due to decreased agricultural production brought on by desertification, it is difficult for impacted people to produce enough food to meet their needs. Increased food insecurity, malnutrition, and dependency on outside food sources are consequences of this condition. Long-term health issues can arise from inadequate availability to nutrient-rich food, particularly in vulnerable populations like children and the elderly.
- ***Migration and Displacement***: Communities may be compelled to migrate in quest of better possibilities if land becomes degraded and incapable of supporting subsistence. Migration brought on by desertification can put a strain on social structures, raise competition for resources, and spark civil unrest. Displaced populations struggle to adjust to new circumstances and risk losing their cultural identity and sense of community (Weltz, 2019).
- ***Economic Losses and Poverty***: Because of decreased agricultural output, deteriorated rangelands, and decreased availability of natural resources, desertification has a detrimental impact on the local economy. These elements lead to fewer income prospects, which feeds the vicious cycle of poverty and instability in the economy. Communities that are affected frequently have few job opportunities, little access to markets, and a greater reliance on outside assistance and support.
- ***Political Unrest and Conflict***: Conflicts over diminishing resources caused by desertification can arise in areas that are

strongly reliant on natural resources. Political instability is exacerbated by disputes over water rights, grazing grounds, and fertile soil, which also impedes efforts at sustainable development. The wide-ranging effects of these conflicts include social discontent, evictions, and even regional destabilization.

The socioeconomic effects of desertification on food security, livelihoods, economic stability, and social dynamics are extensive and diverse. These effects highlight the critical need for efficient policies and actions to combat desertification, advance sustainable land management techniques, and aid impacted populations in fostering resilience and adjusting to changing environmental conditions.

8.5 CASE STUDIES

- **Sahara Desert**: The Sahara Desert is a well-known illustration of how desertification affects a large area. Countries like Niger, Chad, and Mali struggle to maintain their agricultural systems and adjust to the changing environment as arid areas encroach on productive lands. Implementing sustainable land management techniques, reforestation programs, and the promotion of alternative livelihoods like ecotourism have all been used to combat desertification in the Sahara (Runyan, 2013).

  The "Great Green Wall" effort in Niger seeks to stop desertification and repair damaged lands. A belt of trees will be planted on the southern edge of the Sahara Desert as part of this program, offering many advantages such as soil stability, water management, and job possibilities for nearby communities. To enhance soil fertility and raise agricultural output, the initiative also uses sustainable agriculture techniques, such as agroforestry.

  The "Grassroots Women's Initiative for Sustainable Development" in Chad aims to empower women by teaching them how to manage their land sustainably to prevent desertification. The program encourages practices including soil conservation, water harvesting, and agroecology, giving women the tools, they need to rehabilitate degraded lands, increase food security, and make money from sustainable agricultural endeavors.

  The "Great Green Wall for the Sahara and Sahel Initiative" in Mali aims to improve livelihoods, rehabilitate degraded lands, and safeguard biodiversity. The project entails restoring wetlands, grasslands, and forests in addition to promoting sustainable farming methods and animal management techniques. In addition to preventing desertification, these initiatives support ecosystem preservation and adaptation to climate change (Runyan, 2013).

Additionally, ecotourism activities in the Sahara region offer residents alternate means of subsistence. These programs emphasize environmentally conscious tourism behaviors that also help the preservation of our natural and cultural heritage. Communities in the Sahara Desert can lessen their reliance on agriculture and increase their resistance to desertification by diversifying their sources of income.

- **The Gobi Desert**: An alarming rate of desertification has occurred in Asia's Gobi Desert. Desertification is a result of unsustainable grazing methods and climate change, which is endangering pastoral communities and their way of life and accelerating environmental deterioration. Rotational grazing, water-saving methods, and forestry initiatives have all been put into place to combat desertification in the Gobi Desert to restore degraded land and save vulnerable ecosystems (Weltz, 2019).

The "Gobi Desert Initiative" in Mongolia emphasizes community involvement and environmentally sound rangeland management. This effort encourages the use of rotational grazing methods, which entail moving livestock herds periodically to avoid overgrazing and permit the growth of new vegetation. To lessen pressure on the delicate ecology, it also promotes the creation of community-managed conservation areas and the growth of alternative income-generating ventures like eco-tourism and handicrafts.

To increase water efficiency in agricultural activities, further water conservation strategies have been developed. These include the installation of small-scale water harvesting structures and the use of drip irrigation. These actions enhance sustainable agriculture, protect water resources, and lessen communities' susceptibility to droughts and water scarcity.

In the Gobi Desert, reforestation initiatives seek to restore flora and stop soil erosion. Sand dunes may be stabilized, soil quality can be improved, and biodiversity can be increased by using native plants and drought-resistant tree species. These initiatives not only lessen desertification but also aid in carbon sequestration, which lessens the effects of climate change.

- **Thar Desert**: The socio-economic effects of desertification are best illustrated by the Thar Desert in India. Water shortage, low agricultural production, and high levels of poverty are problems in the area. Afforestation, rainwater collecting, and sustainable land management techniques are some of the methods used to stop desertification in the Thar Desert. Additionally, the promotion of alternative occupations like tourism and handicrafts has given local communities economic opportunities while easing the strain on the environment.

  Rainfall harvesting methods have been widely used in the Thar Desert to collect and store rainfall for home and agricultural usage. Building check dams, gathering rainwater from rooftops, and erecting underground storagee tanks are all tried-and-true techniques for replenishing groundwater levels and enhancing water supply during dry spells. These methods assist lessen the effects of droughts while simultaneously addressing the issue of water scarcity (Runyan, 2013).

  Initiatives to reforest the Thar Desert seek to stop soil erosion and restore the natural equilibriumNative tree species that are suitable to arid conditions are planted in order to protect soil moisture, maintain the stability of sand dunes, and provide shade and habitat for wildlife. Participation of the local community in tree-planting projects supports the long-term success of these initiatives and creates more options for sustainable forest management-based livelihoods.

  The Thar Desert employs sustainable land management approaches to increase soil fertility, retain moisture, and increase agricultural productivity. These activities include contour bunding, agroforestry, and soil conservation techniques. To maintain soil health and lessen dependency on chemical inputs, farmers are encouraged to adopt organic farming practices, utilize organic fertilizers, and engage in crop rotation. These methods advance ecological equilibrium, strengthen drought resistance, and support sustainable agriculture.

  The Thar Desert has seen the encouragement of alternative livelihoods to lessen poverty and ease the strain on natural resources. Producing handicrafts like ceramics, rugs, and needlework has grown to be an important source of revenue for regional people. Additionally, the region's distinctive natural landscapes and cultural legacy have drawn travelers, encouraging the creation of sustainable tourism programs that boost the local economy while maintaining the desert ecology (Weltz, 2019).

These case studies demonstrate the many issues and strategies used to manage desertification in various parts of the world. It is possible to stop desertification, safeguard ecosystems, and improve the welfare of communities impacted by this environmental problem by comprehending the specific dynamics and putting context-specific policies into practice.

8.6 LESSONS LEARNT FROM EFFECTIVE RESTORATION PROJECTS

Several effective restoration initiatives offer crucial insights into halting desertification. The Loess Plateau restoration project in China, the Al Baydha Project in Saudi Arabia, and the Great Green Wall effort in Africa all serve as examples of how sustainable land management techniques, reforestation, and community involvement may be used to restore damaged landscapes. These initiatives have shown the value of comprehensive strategies that incorporate ecological, social, and economic facets of land restoration.

Through the planting of a continuous strip of trees throughout the Sahel region of Africa, the Great Green Wall effort seeks to stop the spread of desertification. The project encourages environmental restoration, increases food security, and enriches livelihoods by integrating local communities in tree planting, sustainable land management, and income-generating activities.

Terracing, reforestation, and soil conservation techniques were used to successfully recover damaged land and improve water supplies as part of the Loess Plateau restoration project in China. The project's integrated strategy included involving neighborhood communities, offering technical assistance, and advocating for sustainable agricultural methods (Runyan, 2013).

The *Al Baydha* Project in Saudi Arabia aims to improve livestock management and restore rangelands in desert areas. The project has been successful in restoring degraded regions, improving vegetation cover, and boosting the resilience of pastoral communities by introducing rotational grazing, water conservation measures, and reseeding initiatives.

These restoration initiatives provide insightful insights into the global fight against desertification. They place a strong emphasis on the value of community involvement, the blending of traditional knowledge with scientific methodologies, and the significance of stakeholder collaboration and long-term commitment. It is possible to make substantial advancements in the fight against desertification and the promotion of sustainable land management on a global scale by putting these lessons into practice and scaling up effective techniques.

8.6.1 Observation and Evaluation of Desertification

- ***Techniques Using Remote Sensing and GIS***: Monitoring and evaluating desertification heavily relies on geographic information systems (GIS) and remote sensing technologies. It is possible to track changes in plant cover, soil moisture, and land use patterns across time using satellite photography and remote sensing techniques. GIS tools improve decision-making processes by assessing and mapping the amount of desertification, identifying sensitive areas, and collecting data. These technologies make it possible for academics and land managers to acquire precise and current information on the processes and patterns underlying desertification, permitting focused interventions and efficient resource allocation (Weltz, 2019).
- ***Approach to Land Degradation Neutrality (LDN)***: A framework for evaluating and tracking the balance between land degradation and restoration efforts is provided by the Land Degradation Neutrality (LDN) concept. The goal of LDN is to prevent additional desertification by achieving a state in which the quantity and quality of land resources are stable or improve over time. The LDN strategy enables governments to track their progress in tackling desertification and execute necessary steps to attain land degradation neutrality by monitoring land degradation indicators and setting targets.
- ***Models and Indices of Desertification***: To quantify and assess the processes of desertification, numerous indices, and models have been created. These instruments evaluate the degree of desertification and identify the most pressing locations for

restoration and intervention by taking into account variables including plant cover, soil erosion rates, rainfall patterns, and changes in land use. Indicators of vegetation health and production, which are crucial signs of desertification, include the Normalized Difference Vegetation Index (NDVI) and the Soil Adjusted Vegetation Index (SAVI). Models that forecast soil erosion rates and help identify regions at high risk of desertification include the Soil Water Assessment Tool (SWAT) and the Revised Universal Soil Loss Equation (RUSLE).

- ***Local and International Monitoring Projects***: Monitoring and reversing desertification are the focus of numerous regional and international projects. The United Nations Convention to Combat Desertification (UNCCD) is crucial in organizing global efforts, assisting member nations with the implementation of monitoring systems, and exchanging best practices. Through the use of remote sensing data, regional programs like the African Monitoring of the Environment for Sustainable Development (AMESD), the Global Observation of Forest and Land Cover Dynamics (GOFC-GOLD) initiative, and the Climate Change Initiative (CCI) of the European Space Agency all contribute to monitoring desertification and providing useful information for policy-making and sustainable land management (Runyan, 2013).

Understanding the dynamics of desertification, monitoring the progression of the phenomenon's eradication, and formulating successful intervention measures all depend on these monitoring and evaluation techniques in combination with international cooperation. Decision-makers and land managers can implement targeted activities to combat desertification and advance sustainable land use practices by using these tools and initiatives.

8.7 ALTERNATIVES TO DESERTIFICATION

8.7.1 Ecologically Sound Land Management Techniques

- ***Agriculture in Conservation***: Techniques used in conservation agriculture, such as minimal tillage, crop rotation, and cover crops, encourage the preservation of soil, the retention of water, and biodiversity. These methods increase soil fertility, reduce soil erosion, and increase agricultural output in arid areas. Conservation agriculture preserves soil moisture, lowers evaporation, and guards against wind and water erosion by limiting soil disturbance and keeping a continuous ground cover (Weltz, 2019).

- ***Agroforestry and Forestry***: Agroforestry systems, which include trees with crops or cattle, can repair degraded areas, improve the retention of soil moisture, and offer a variety of advantages, including lumber, fuelwood, and fodder. Trees help generate microclimates that reduce water loss through evaporation and offer shade for livestock and crops. To improve the resilience of ecosystems, stabilize the soil, and preserve water, afforestation projects entail planting trees in arid regions. Afforestation programs decrease soil erosion, enhance soil structure, and foster biodiversity by restoring tree cover.
- ***Management of Watersheds***: Terracing, contour plowing, and the installation of check dams are examples of integrated watershed management techniques that assist retain water and lessen soil erosion. These methods are intended to reduce water flow so that it can seep into the earth and rehydrate the groundwater. Water harvesting strategies like creating ponds or reservoirs and restoring natural water bodies can increase the amount of water available for home and agricultural uses. Additionally, community involvement and local soil and water conservation measures are part of watershed management (Runyan, 2013).
- ***Management of Rangelands and Pasturelands***: Through rotational grazing, rest intervals, and range enhancement strategies, rangelands and pasturelands can be managed sustainably to avoid overgrazing, preserve vegetation cover, and advance ecosystem health. To give the plants time to recuperate, rotational grazing includes moving livestock between several grazing areas. Plants can regrow during periods of rest, preventing their extinction and fostering biodiversity. Techniques for enhancing a range include planting new seeds, controlling erosion, and constructing windbreaks. Enhancing livestock productivity, preserving traditional livelihoods, and maintaining ecological services are all benefits of restoring damaged rangelands.

Communities and governments may lessen the effects of desertification, rehabilitate damaged landscapes, and foster long-term resilience by putting these sustainable land management techniques into practice. These solutions help with sustainable development, food security, and the preservation of natural resources in addition to addressing the immediate problems caused by desertification.

8.7.2 Management of Integrated Water Resources

- ***Techniques for Harvesting Rainwater and Irrigation***: Check dams, contour bunding, and other methods for capturing and storing rainwater can replenish groundwater aquifers and supply more water for agriculture. By collecting and storing rainwater flow from roofs or other surfaces, rainwater harvesting devices lessen the need for outside water sources. By creating small earthen embankments along the land's contours, contour bunding can catch and slow down rainwater, allowing it to permeate the soil and replenish groundwater. Crossing river channels, check dams are built to impound water and create tiny reservoirs for agriculture. Water is delivered directly to the plant root zone via efficient irrigation techniques like drip irrigation and precision agriculture, which minimize water loss due to evaporation and guarantee optimal water use efficiency (Weltz, 2019).
- ***Desalination and Recycling of Water***: Desalination technology can offer a reliable source of water for drinking and cultivation in areas with few freshwater resources. Desalination procedures clean saltwater or brackish water of salt and other contaminants to create freshwater fit for a variety of uses. Treatment and reuse of wastewater for irrigation, industrial processes, or non-potable domestic purposes are part of water recycling and reuse activities. These methods assist relieve pressure on water availability in arid locations by lowering demand for freshwater sources (Runyan, 2013).
- ***Water-Savvy Farming Techniques***: Combating desertification and maintaining sustainable water use in agriculture depend on using water-efficient farming techniques. Using crop varieties that can withstand drought and thrive in dry environments lowers water requirements while preserving agricultural productivity. Through practices like mulching, conservation tillage, and soil amendments, it is possible to improve soil moisture conservation and lower water loss through evaporation. Water waste is reduced via precision irrigation systems that give the appropriate quantity

of water to crops depending on their unique demands, such as soil moisture sensors and computerized irrigation scheduling. By using these water-saving agricultural techniques, you may improve crop yield while lowering the risk of soil erosion and desertification.

8.7.3 Policy Interventions and Global Collaboration

- ***Conventions and Treaties Regarding Desertification***: Nations can work together to tackle desertification via international agreements like the United Nations Convention to Tackle Desertification (UNCCD). These protocols make it easier to share best practices, technologies, and expertise to address the underlying causes of desertification. Promoting sustainable land management and land restoration requires national policies and laws that are in compliance with these international agreements. To effectively address desertification, governments and stakeholders must give top priority to implementing laws that encourage sustainable land use, provide incentives for conservation efforts, and control access to land and water.
- ***Financial Incentives and Support Systems***: Sustainable land management techniques can be promoted by giving farmers, landowners, and communities financial incentives, subsidies, and grants. Financial assistance can reduce the upfront expenses related to putting technology, training plans, and infrastructure development required to stop desertification into action. Establishing financial mechanisms that give priority to initiatives to mitigate and restore desertification is something that governments, international organizations, and private organizations can work on together. These monetary rewards are essential for encouraging sustainable behavior and facilitating the switch to more resilient land management systems (Weltz, 2019).
- ***Knowledge Exchange and Resource Development***: To spread best practices, develop local expertise, and increase awareness about desertification, it is crucial to promote knowledge-sharing platforms and capacity-building programs. Through workshops, training programs, and educational campaigns, communities are given the tools they need to actively participate in sustainable land management and restoration initiatives. These tools include scientific research, traditional wisdom, and success

stories. Training programs on organic agricultural methods, soil conservation techniques, and water management techniques are examples of capacity-building projects. Communities are better able to address the problems caused by desertification and contribute to long-term solutions by building local capacity. Effective information sharing networks and capacity-building initiatives depend on cooperation between governments, research institutions, and civil society organizations.

8.7.4 Strategies for Adapting to Climate Change

- ***Climate-smart Agriculture***: Putting climate-smart farming practices into place is essential for coping with the problems brought on by climate change. Agroecology, precision farming, and the cultivation of adaptable crop varieties are just a few of the practices that make up climate-smart agriculture. To increase agricultural output while reducing its negative effects on the environment, agroecology emphasizes ecological concepts such as crop variety, organic pest management, and soil protection. Agroecology develops ecosystem resilience and improves the adaptability of farming systems by fostering biodiversity and lowering dependency on chemical inputs (Runyan, 2013).

  Depending on the crop and soil conditions, precision farming makes use of technology and data to maximize the use of resources, such as water and fertilizers. Farmers are better able to adapt to changing climatic circumstances by using precision agriculture techniques to decrease inputs, reduce waste, and boost efficiency.

  To ensure food security, it is crucial to cultivate resilient crop types that are better adaptable to shifting climatic circumstances. These types have characteristics that allow them to grow in harsh settings, including heat tolerance, drought resistance, disease resilience, and quicker maturation times. Farmers that use these crops in their farming operations are better equipped to endure the effects of climate change and maintain consistent harvests.

  Soil health and conservation are also given top priority in climate-smart agricultural practices. Farmers can optimize soil structure, and boost water-holding capacity and nutrient availability by implementing measures including conservation tillage, cover crops, and organic matter incorporation. Farming systems are more resistant to droughts and extreme rainfall events when the soil is healthy because healthy soils are better able to absorb and store water.

Adopting climate-smart agricultural techniques supports ecosystem health, greenhouse gas emissions reduction, and sustainable food and livelihood production. Communities can adapt to the changing climate, guarantee food security, and promote long-term resilience in agricultural systems by putting these measures into practice (Weltz, 2019).

- ***Green Technologies and Renewable Energy***: To counteract desertification, switching to renewable energy sources is a crucial part of adaptation methods to climate change. A sustainable alternative to fossil fuels, renewable energy sources including solar, wind, and hydropower help to reduce greenhouse gas emissions and combat climate change. Countries can indirectly prevent desertification by minimizing environmental degradation and lowering carbon emissions by moving away from fossil fuel-based energy production.

  In arid and semi-arid areas, solar energy is a particularly attractive renewable energy source. Since sunshine is frequently abundant in these regions, solar energy is a practical and effective energy source. To capture solar energy and turn it into power, solar panels can be put on roofs, farms, or open areas. Communities may help preserve ecosystems and landscapes at risk of desertification by increasing the use of solar energy. This will minimize their carbon footprint and lessen their reliance on non-renewable energy sources (Runyan, 2013).

  Another clean and sustainable energy source that can aid in the fight against desertification is wind power. Wind turbines use the wind's energy to produce electricity. To provide sustainable energy and lessen the dependency on fossil fuel-based power plants, wind farms can be built in windy areas, such as coastal regions and wide plains. Communities can reduce climate change and aid in the preservation of natural resources by using wind power.

  The adoption of green technologies, together with the use of renewable energy sources, is essential for halting desertification and fostering adaptation to climate change. Environmentally friendly techniques and developments that reduce resource use, waste production, and environmental impact are referred to as "green technologies." Examples include smart grid systems, electric automobiles, and environmentally friendly waste management practices.

The use of green technologies and renewable energy has many advantages in the fight against desertification. These measures assist in reducing greenhouse gas emissions, which lessens climate change, a primary contributor to desertification. Additionally, the use of green technologies and the transition to renewable energy sources support economic resilience, promote sustainable development, and create green jobs.

Governments, corporations, and people all have a part to play in promoting green technologies and renewable energy. Technology advances, financial incentives, and policy assistance are essential for hastening the shift to a low-carbon economy and supporting sustainable practices. Communities can adapt to climate change, save ecosystems, and support international efforts to mitigate desertification by embracing renewable energy and green technologies (Weltz, 2019).

- ***Restoration and Conservation of Ecosystems***: Ecosystem restoration and conservation is a key method for coping with climate change that also significantly contributes to the fight against desertification. Ecosystems including forests, wetlands, and mangroves offer crucial functions that support the general wellbeing and adaptability of landscapes. These ecosystems have several advantages in terms of carbon sequestration, water control, soil stabilization, and habitat restoration in addition to supporting biodiversity.

  In terms of climate change adaptation and ecosystem restoration, forests are very important. They serve as carbon sinks, drawing huge amounts of carbon dioxide from the atmosphere and storing it there. Planting new trees, recovering damaged forest regions, and putting sustainable forestry techniques into practice are all parts of forest restoration programs. These initiatives support soil conservation, stop erosion, and safeguard watersheds in addition to helping to battle climate change. The habitats that restored forests provide for various plant and animal species to increase biodiversity (Runyan, 2013).

Peatlands, marshes, and other types of wetlands are essential ecosystems for preventing desertification and adjusting to climate change. Wetlands serve as organic water filters that clean the water and stop soil erosion. Additionally, they act as carbon sinks, storing large amounts of carbon and lowering greenhouse gas emissions. Water availability can be increased, water flow can be controlled, and ecosystems' overall resilience can be increased by protecting and restoring wetlands.

Mangroves are coastal ecosystems that are essential for reducing the effects of extreme weather events and adapting to climate change. Natural barriers like mangroves lessen the effects of storm surges and coastal erosion. They also support nearby fishing villages and serve as habitats for a variety of marine organisms. Mangrove preservation and restoration can improve coastal resilience, protect biodiversity, and support sustainable livelihoods.

Combating desertification and climate change also requires protecting the current ecosystems. Protected areas, national parks, and nature reserves support the maintenance of biodiversity, the preservation of essential habitats, and the protection of ecosystem services. Implementing efficient management techniques, reducing the negative effects on humans, and promoting sustainable resource use are all part of conservation initiatives (Runyan, 2013).

Initiatives for ecosystem restoration and protection must actively involve the community. Long-term success is ensured by involving local communities, indigenous peoples, and stakeholders in decision-making processes and using sustainable techniques. Communities can improve the resilience of landscapes, mitigate climate change, and stop the spread of arid regions by restoring and protecting ecosystems.

The promotion of ecosystem restoration and conservation depends heavily on governmental assistance, legislative frameworks, and international cooperation. Restoration initiatives and conservation measures can be more easily put into action with the help of financial incentives, land-use restrictions, and capacity-building initiatives. Countries, organizations, and communities must work together to maintain and restore ecosystems to prevent desertification and adapt to climate change (Weltz, 2019).

Chapter

9 ECOLOGICAL IMPACTS OF CLIMATE CHANGE

CONTENTS

9.1 INTRODUCTION

Climate change's ecological implications are rarely considered in studies on natural disasters and risks. The word "ecological impacts" in this context refers to changes in organism abundance, distribution, and functions, which include interactions and environmental consequences. When seen from the perspective of the organisms, the bulk of the outcomes may be hazardous to humans and potentially have disaster-scale ramifications.

The distribution of various flora types on the earth is influenced by climate. Furthermore, each species has its distribution patterns, the limitations of which are primarily determined by global climate. These temperature distribution patterns demonstrate how temperature influences ecological activities like competition, pollination, and herbivory, as well as plant survival, physiology, and growth varied plant species are suited to varied climatic conditions, ranging from cold-tolerant, slowly growing alpine plants to rapidly growing trees in the tropical tropics. To precisely pinpoint the ecological consequences of climate change, both the capacity to perceive changes in an ecosystem and the ability to definitively correlate those changes to a change in temperature are necessary. Any change in an environmental variable must be identified, which necessitates the use of a trustworthy dataset including time-lapsed measurements recorded repeatedly (Feng, 2015).

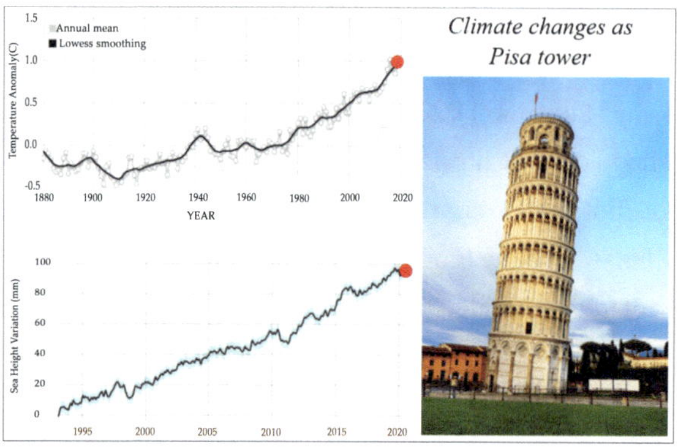

Figure 9.1: Global temperature and sea level rising (**Left**) and Pisa tower as a metaphor (**Right**).

Source: Glasnovic Z, Margeta K, Logar NZ. Humanity Can Still Stop Climate Change by Implementing a New International Climate Agreement and Applying Radical New Technology. *Energies*. 2020; 13(24):6703. https://doi.org/10.3390/en13246703 [accessed 12 Jun, 2023]

The effects of climate change on plants are complex, and a species' existence or absence in a specific location is influenced by variables other than its ability to tolerate local climatic circumstances. The impact of temperature on a plant's ability to compete with other species defines climatic restrictions in many cases. Climate change may also impede connections between plants and their pollinators, mycorrhizae, herbivores, or pathogens. Rising temperatures are the most well-known impacts of climate change, but in the long run, less predictable effects such as changes in precipitation or infrequent severe occurrences may be more significant. Changes in atmospheric composition, particularly carbon dioxide levels, can influence plant relationships and performance. Plants are adjusting to climate change through changed phenology and distribution patterns, with species usually migrating toward cooler places (Williamson, 2013).

Interactions and impacts on the abiotic environment are examples of biological functions. When we discuss ecological implications, we are referring to changes in the quantity, distribution, and functions of species. When seen from the perspective of the organisms, the bulk of the repercussions may be detrimental to mankind and have disaster-scale ramifications.

Environmental responses to climate change were well understood long before global warming became a concern. Early environmental studies recognized the link between organisms and climate and debates on the implications of climate change began in the eighteenth century when the historical climatic change was known. When glacial geomorphology was initially used to infer former climates, ecological data were employed to increase our knowledge of climatic change, and this work is currently ongoing (Feng,. 2015)

The most visible response to broad climate change is a shift in an organism's geographic range. Moving permits organisms to dwell in conditions that they have adapted to. As a result, in the 20,000 years since the Last Glacial Maximum (LGM), we have seen changes in the poleward distribution of species over their entire ranges. Animals adapt via changes in their physiology, behavior, and life histories that may be impacted by natural selection and even modify their genomes when they move across ranges and during shorter periods. The most significant of these reactions is phenology change, such as spring greening in North America.

According to several studies, the length of the vegetative growth season has increased as the climate has changed. This tendency has been more prominent at higher latitudes when the regional temperature has increased the most. A comprehensive examination of the growing season in gardens around Europe indicated that spring activities, such as leaf unfolding, occur 6 days sooner and fall events, such as leaf coloring, occur 4.8 days later. According to this study, the length of the growing season rose by 10.8 days per year on average between 1959 and 1993. A 61-year record of phenophases in southern Wisconsin (43°N), including migratory bird arrival dates and dates of spring flowers' first bloom, found an increase in earliness of 0.12 days each year, or 7.3 days throughout the research. Some climate change reactions offer dangers to species because they can increase the possibility of extinction by reducing populations or fragmenting them geographically. One of the main ideas is that when species ranges migrate poleward in response to climatic warming, the area in any climate must reduce. When this is combined with the most dependable connection in biogeography, which claims that the number of species is directly connected to the area, the projected consequence is species loss (Williamson, 2013).

An extinction debt or a postponement of the inevitable final extinctions, according to studies, may be the cause of this pattern. However, the species under consideration do not expect extinction, and their reactions, except for some animal migrations, are neither directed nor planned. From the perspective of humans, two elements should be considered. First, as animals, humans are vulnerable to danger as other organisms adapt to climate change. Other chapters in this book look at changes in live disease vectors, whereas other volumes in this series look at changes in other environmental elements that have a direct impact on humans, such as agriculture. Here, we focus on how climate change may affect ecosystem services (Feng, 2015).

Climate variability and extremes have the potential to create rapid and lasting changes in ecosystems. Because severe events are chaotic and rarely predictable, abrupt changes in ecological systems (ACES) are difficult to track experimentally. The authors, however, strongly advocate scientists prioritize recognizing, analyzing, and planning for ACES in response to climate change. The intensity and frequency of certain drivers are changing at an increasing rate; no «new normal» (equilibrium) exists. The analysis shows considerable generalizations, which prompt questions and ideas for additional investigation. These are some examples: Some ecological system characteristics are more vulnerable to abrupt change than others; climate extremes may be more likely than mean trends to cause abrupt change (for example, coral bleaching is caused by intense heatwaves rather than gradual ocean warming); multiple drivers frequently interact to produce ACES (for example, climate change-driven drought and extreme fire can lead to abrupt changes of terrestrial ecosystems from forest to non-forest, introduced pathogens can lead to abrupt changes of pathogen

The features of belowground ecological groupings that make them resistant, resilient, or vulnerable to climate extremes are summarized. They also consider historical circumstances and climatic extremes. Soil microbial communities play an important role in the biogeochemical cycle.

Figure 9.2: Social-ecological resilience.

Source: During R, Van Assche K, Van Dam R. Relating Social and Ecological Resilience: Dutch Citizen's Initiatives for Biodiversity. *Sustainability*. 2022; 14(7):3857. https://doi.org/10.3390/su14073857 [accessed 12 Jun, 2023]

Life-history strategy (growth rate, resource usage efficiency) and microbial food web variation (fast and slow energy pathways found in bacterial vs. fungal food webs) are essential inherent properties of these communities that contribute to their resilience. While slow energy channels (such as fungi) cycle nutrients slowly and attenuate reactions to perturbations, conferring resilience to change, rapid energy channels (such as bacteria in the soil) recycle nutrients quickly and recover quickly from disturbances.

Changing the relative significance of both routes, on the other hand, may cause an ecosystem to become unstable. The complementary activities of these two energy pathways can aid an ecosystem's recovery from perturbations. Extrinsic qualities include environmental unpredictability and the plant community's contributions to soil carbon, moisture, and nutrients. Although the researchers acknowledge that the response of subsurface communities to long-term stress is well documented, they highlight the prospect of sudden ecological change and the response to climatic extremes as critical knowledge gaps that should be filled through experimental study (Williamson, 2013).

Longer-term perspectives are required to better understand how ecological communities adapt to change. Researchers use palaeoenvironmental records of pollen and charcoal from temperate forests in the Northern and Southern Hemispheres to investigate the influence of fire on the tree species composition of temperate forests. They found that local characteristics such as temperature, soil quality, and historical legacies influence how adaptable or sensitive forest species composition is to change fire regimes. Extreme events paired with biophysical feedback can lead ecosystems composed of long-lived species to shift composition in response to a single fire event in some situations. During the Late Quaternary, temperate forests underwent both long periods of stability and dramatic shifts in reaction to climatic change and human actions (burning for land clearance). To anticipate future ecological scenarios under extraordinary Holocene and post-Holocene warming rates, a site-specific knowledge of stability vs disequilibrium is essential. Climate change reduces terrestrial biodiversity and has an influence on ecosystem carbon storage both directly and indirectly through changes in land use, such as the development of agriculture as a result of climate change (Feng, 2015).

9.2 OCEAN ACIDIFICATION

Climate change causes ocean acidification. Ocean acidification happens when seawater and atmospheric Carbon dioxide react chemically. Water interacts with dissociated Carbon dioxide. Carbon dioxide levels have increased absorption. The pH decreases, making the water more corrosive. Ocean acidification is hypothesized to be produced by atmospheric carbon dioxide, largely from fossil fuel combustion. Carbon dioxide absorption acidifies the waters. Some marine species cannot survive due to ocean acidification. Due to increased atmospheric carbon dioxide, oceans are absorbing a lot of it. Since the 19th century, the oceans have absorbed 33% of human-caused Carbon dioxide. The event raised ocean acidity by 30%. If carbon dioxide emissions continue, ocean pH may increase. Ocean acidification will increase. Reducing carbon dioxide emissions may decrease ocean acidification best. Ecosystems can adapt to ocean acidification by managing stress (Williamson, 2013).

9.3 SPECIES EXTINCTION

The phenomenon of climate change is identified as a contributing factor to the extinction of various species. Certain species are at risk of extinction due to their inability to adjust to alterations in their surroundings. Species that are cold-blooded and inhabit regions with high latitudes and altitudes tend to exhibit accelerated growth rates to offset the limited duration of their growing season. Conversely, marine invertebrates thrive under conditions that are optimal for their growth. Elevated temperatures above the optimal range have been observed to stimulate metabolic processes and decrease body mass, even in the presence of heightened food consumption. This phenomenon may heighten the susceptibility of organisms to predation. Even slight fluctuations in temperature can negatively impact the growth efficiency and survival rate of rainbow trout.

Numerous species of aquatic flora and fauna depend on glacially-fed water sources to thrive in frigid aquatic habitats. The cutthroat trout is among several species of freshwater fish that necessitate low-temperature water for their growth and reproduction. The decline in glacial production could potentially result in insufficient stream flow, which may impede the flourishing of various species. Within the cold waters, a crucial organism referred to as oceanic krill thrives and functions as the primary sustenance for marine mammals, notably the blue whale. The potential impact of augmented freshwater input from glacial runoff and anticipated alterations in the thermocline circulation of global oceans on fisheries that currently serve as a source of sustenance for human populations cannot be ignored.

As per a research study, approximately 14% of terrestrial species are currently facing the threat of extinction. Inevitably, if the global temperature increases by 1.5 degrees Celsius, it is highly probable that approximately 14% of all land-based flora and fauna will face the risk of extinction. If temperatures increase significantly, the issue is likely to exacerbate. For instance, a rise of 3 degrees Celsius in temperature could potentially lead to the extinction of as much as 29% of species that inhabit the land. According to experts, certain flora and fauna, particularly those situated in Polar Regions, may encounter temperatures that surpass their past climatic encounters in the forthcoming decades. The authors emphasized the vulnerability of numerous ecosystems to climate extremes such as heatwaves and droughts, even with a mere 1.2 degrees Celsius of warming (Feng, 2015).

As per the evaluation, a considerable proportion of approximately 47% of the species have experienced a reduction in their population due to the impact of climate change. The phenomenon of global warming has led to the eradication of various local populations of multiple species. The American pika has experienced a considerable reduction in its past habitat within the Sierra Nevada Mountains of California, plausibly due to the effects of climate change. Due to their adaptation to cold environments, these diminutive organisms are particularly vulnerable to exceedingly elevated temperatures.

Should the current pace of climate change persist, it is projected that approximately one-sixth of all species may face extinction. The regions of South America, Australia, and New Zealand have been identified as highly susceptible to various forms of vulnerability. Familiarity with these patterns can aid in our readiness to mitigate and potentially avert the depletion of biodiversity resulting from the impacts of climate change (Williamson, C. E. 2013).

9.4 RISK OF INVASIVE SPECIES

The relationship between climate change and the spread of invasive species remains unclear. The introduction of invasive species can lead to the displacement of native species and disturbance of ecological processes as they encounter difficulties in adapting to the new environment. Invasive alien species (IAS) refer to non-native organisms, including animals, plants, and other life forms, that are introduced into an ecosystem and have a detrimental effect on the local biodiversity, ecosystem services, or human well-being. The activities of invasive alien species (IAS) have been found to exert a considerable impact on the extinction of various species and the reduction of biodiversity. Furthermore, they represent a potential hazard to worldwide food production (Feng, 2015).

The impact of invasive alien species (IAS) is exacerbated by climate change. The facilitation of the establishment and propagation of various alien species is observed to be more feasible, and the probability of their invasiveness is amplified by the phenomenon of climate change. The adverse impacts of climate change, such as hurricanes, floods, and droughts, have the potential to facilitate the spread of IAS to novel habitats and undermine the resilience of ecosystems. Furthermore, the impacts of climate change give rise to novel pathways for the introduction of invasive alien species. Numerous IAS exhibit a remarkable ability to swiftly migrate towards higher latitudes and elevations in response to the warming climate, thereby surpassing the performance of indigenous species. The phenomenon of climate change has the potential to create favorable conditions for the successful colonization of non-native species that have been repeatedly introduced by humans but have not yet been able to establish themselves, thereby resulting in the emergence of novel invasions. With increasing temperatures, certain ecosystems, such as temperate forests and freshwater systems, may become more conducive to the colonization of IAS due to the removal of thermal barriers (Feng, 2015).

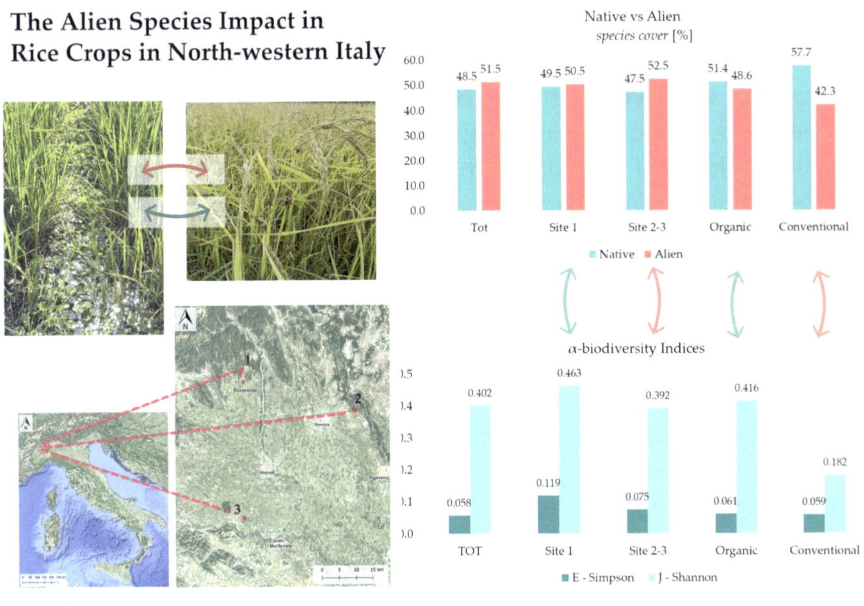

Figure 9.3: Regional invasive species and climate change-an example from Italy.

Source: Vagge I, Chiaffarelli G. The Alien Plant Species Impact in Rice Crops in Northwestern Italy. *Plants*. 2023; 12(10):2012. https://doi.org/10.3390/plants12102012

The regional expansion of IAS is expected to have diverse social and environmental consequences. The phenomenon of biological invasions presents a notable threat to worldwide food security and lifestyle, particularly in developing countries. As a result of their dependence on subsistence and smallholder agriculture, these countries generally exhibit a limited ability to prevent and manage biological invasions.

The introduction of IAS has been found to diminish the resilience of natural ecosystems, rendering them more vulnerable to the impacts of climate change. Certain species of grasses and trees belonging to the IAS category have the potential to considerably alter fire patterns, particularly in areas that are experiencing an increase in temperature and aridity. As a result, the incidence and severity of wildfires are on the rise, thereby endangering ecosystems, societies, and human well-being. The impact of IAS on the welfare of plants and animals could potentially result in negative consequences for agricultural systems (Williamson, 2013).

Considering the interrelationships between climate change and IAS, policy responses to these challenges must take into account the exacerbating effects of the former on the latter. The integration of IAS into climate change policy can be achieved by implementing measures for the prevention and control of IAS, as well as ensuring that climate change initiatives do not aggravate the issue of IAS. The explicit integration of climate change into risk assessments of IAS is recommended to facilitate the identification of potential future hazards.

The prioritization of ecosystems according to their vulnerability to climate change and IAS would enable the formulation of measures to preclude their introduction. To proactively address the issue of invasive alien species, it is imperative to develop efficient biosecurity measures that aim to restrict primary pathways of introduction. Additionally, it is crucial to establish early warning systems and prompt eradication protocols. The imperative of collaborating on a worldwide level is paramount.

9.5 HABITAT LOSS AND FRAGMENTATION

The worldwide deterioration of natural habitats presents a significant threat to various plant and animal species. Notwithstanding the obstacles, there remains potential for the forthcoming due to the considerable influence of habitat depletion on the Earth's ability to sustain life. The devastation of habitats results in adverse effects not only on individual species but also on the entire ecosystem. The term "destruction of habitat" refers to the process of either eliminating or modifying the essential conditions required for the survival of flora and fauna. While acknowledging that other factors may also contribute, it is widely accepted that human activity is the primary cause of habitat destruction. The clearance of land for agricultural, grazing, mining, drilling, and urbanization purposes has a detrimental impact on the species that inhabit forests, which are known to house 80 percent of all species present on the planet. Annually, more than 15 billion trees are felled. As per a research article published in the scientific journal Nature regarding tree density, it has been observed that the worldwide count of trees has declined by 46% since the inception of human civilization (Feng, 2015).

Deforestation not only results in the loss of habitats but also reduces the ability of trees to sequester carbon, a crucial process in mitigating the impacts of climate change.

The state of aquatic ecosystems, including waterways, coastal areas, and the ocean, is alarming. Estuaries and coastal marshes are the primary reproductive habitats for the vast majority of marine organisms. The alteration of inland wetlands through dredging and filling has been observed to harm the reproductive and parental care capabilities of certain species. Streams and rivers are effective conduits for the transportation of pollutants and effluents from terrestrial sources to the ocean, posing a significant threat to the well-being of marine organisms such as fish, wildlife, and vegetation. The process of deforestation, when occurring at a considerable distance from the coastline, can impede the necessary sunlight required for the sustenance of coral reefs. This, in turn, leads to erosion that enters the ocean and deposits detritus in shallow marine areas.

Alterations in the composition of the atmosphere and oceans possess the capacity to exert a noteworthy influence on the biosphere, which constitutes the narrow stratum of existence on the planet that is intimately linked to the atmosphere and hydrosphere and constitutes the fundamental basis for human societies. Hence, modifications to the biosphere that impact specific areas can potentially exert an influence on either the local vicinity or the entirety of the planet. Anthropogenic greenhouse gas emissions are responsible for both climate change and ocean acidification, posing a threat to the resilience and survival of natural ecosystems as well as the human communities that rely on them. The potential hazards associated with these risks can result in significant consequences, as has been increasingly apparent in recent times. The Earth's climate has undergone substantial warming, and unless carbon emission pathways are significantly modified, additional warming is expected (Williamson, 2013).

Despite the widespread deterioration of natural habitats, there are still grounds for a positive outlook. Based on scholarly investigations, it has been suggested that the conservation of approximately 50 percent of the Earth's land and oceans could potentially ensure the survival of various plant and animal species. At present, the proportion of protected land and water stands at 15% and 7% respectively, thereby presenting a formidable yet feasible goal (Feng, 2015).

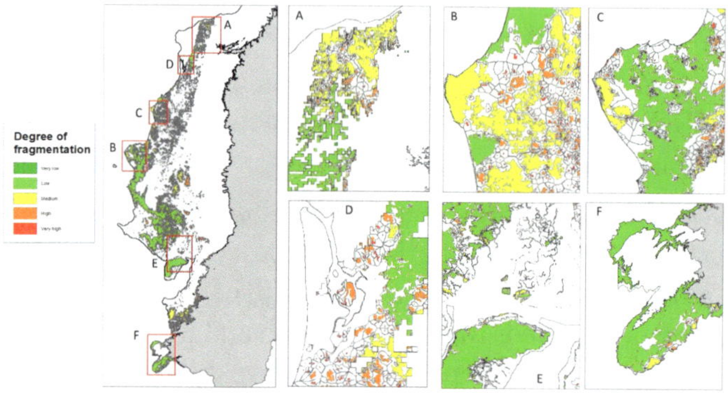

Figure 9.4: Degree of fragmentation of the seasonal dry forest in Ecuador in 1990 (left) using the PFI. The figures in details (**A–F**) show the state of fragmentation of the patches and their area of influence (black lines): (PFI ≥ 0.8), high (0.6 ≤ PFI ≤ 0.79), medium (0.4 ≤ PFI ≤ 0.59), low (0.39 ≤ PFI ≤ 0.2), and very low (PFI ≤ 0.2).

Source: Rivas CA, Guerrero-Casado J, Navarro-Cerrillo RM. A New Combined Index to Assess the Fragmentation Status of a Forest Patch Based on Its Size, Shape Complexity, and Isolation. *Diversity*. 2022; 14(11):896. https://doi.org/10.3390/d14110896 [accessed 12 Jun, 2023]

It is expected that certain climatic anomalies will increase in frequency as a result of climate change, as indicated by scientific research. The aforementioned factors possess the potential to exert a noteworthy adverse influence on biodiversity, especially in instances where meteorological thresholds are surpassed, thereby resulting in the extinction of populations. In the event of persistence, the possibility of populations recovering may become unattainable, ultimately resulting in regional extinctions.

By utilizing comprehensive time series data on butterflies in the United Kingdom, it becomes feasible to measure the population's reactions to past notable droughts, along with the interplay between habitat area and fragmentation. This study integrates present data with future forecasts derived from multiple climate models, considering different Representative Concentration Pathways (RCPs) and accounting for the combined effects of diverse terrain characteristics. It is projected that the extinction of drought-sensitive butterfly populations may transpire as soon as 2050 under the RCP8.5 scenarios, which is associated with emissions resulting from conventional business practices. Through the regulation of landscapes, particularly using inhibiting habitat fragmentation, the likelihood of survival until the middle of the century increases from an almost negligible value to a range of 6-42% (with a 95% confidence interval). It has been observed that the likelihood of persistence is higher than 50%, subject to moderate climate change (RCP 2.6) and semi-natural habitat restoration, until the year 2100. Based on the available data, it can be inferred that the restoration of semi-natural landscapes to mitigate fragmentation is a more efficacious approach towards ensuring the survival of drought-sensitive butterflies, as compared to the mere expansion of their habitat size. However, it is imperative to note that the success of this strategy is contingent upon the implementation of substantial emission reductions (Feng, 2015).

9.6 RANGE SHIFT AND DISTRIBUTIONAL CHANGES

Environmental specializations are shifting as a result of climate change, and animals are modifying their habitat ranges accordingly. Understanding the advantages and disadvantages of these transitions is crucial for ensuring proper biodiversity management. Certain modifications enable the survival of species but may be detrimental to extant species in certain areas. When species reach their biological limits and are affected by climate change, they must modify their environment to persist. In addition to climate change, other human activities such as municipal or infrastructure development and altering land uses in general may have an impact on habitat variability. Because there may not be sufficient evidence to define a species' historical range or because that range may have been dynamic in the recent past, it is difficult to determine a species' habitats and whether or not change is occurring. In addition, climate change will expedite range shifts. However, plant species may alter their latitude or elevation in response to climate change, making range shifts potentially easier to detect. As global temperatures rise, all species can expect their ranges to shift poleward and equatorward. As they can submerge deeper to escape the warmer surface waters, marine species may be able to maintain their equatorial distribution (Williamson, 2013).

As temperatures rise, many North American species' habitat ranges migrate northward and upward. Plants and animals have recently migrated at a rate of 36 feet (0.011 kilometers) per decade to higher elevations in both terrestrial and aquatic environments and at a rate of 10.5 miles (16.9 kilometers) per decade to higher latitudes. While this results in a range expansion for some species, it results in a shift to less hospitable habitat, increased competition, or a range reduction for others, with some species having nowhere to go because they have reached the summit of a mountain or the northernmost point of suitable habitat. These conditions cause local extinctions of both plants and animals in some regions. It is anticipated that by 2100, between 5 and 20 percent of the land in the United States will contain a variety of vegetative biome ranges.

According to the IPCC, species in terrestrial, freshwater, and marine ecosystems are already being impacted by climate change. As global temperatures rise, the consequences will intensify. In addition to more frequent and severe extreme events such as droughts, floods, and fires, the majority of animal populations are threatened by habitat degradation, alterations to the water cycle, and thermal stress. These effects also affect individuals, increasing resource competition among all forms of life.

Consider the following impacts on a few of the highlighted species:

- *Corals in Warm-Water Environments*: Red coral, which thrives in balmy waters, is capable of living for hundreds of years. These species are particularly susceptible to global warming. The Intergovernmental Panel on Climate Change (IPCC) estimates a 70% loss of warm water corals in a future with moderate warming, with temperature rises limited to 1.5°C. When temperatures rise by 2°C or more, virtually all corals in warm water perish (Feng, 2015).
- *A Tall Oak Tree*: There are approximately 500 distinct species of oak trees, many of which can live for over 250 years. Oaks have shifted their geographic range and undergone genetic development to adapt to climate change. However, these species suffer when wildfires become more frequent and intense, parasites have more opportunities to propagate, and droughts become more severe in some regions (Williamson, 2013).
- *Whales*: Bowhead whales may survive in the wild for over 200 years, making them a species with a lengthy lifespan. The effects of climate change on bowhead habitat utilization, dispersion, and migration timing are evident.
- *Nature Contributes to the Solution*: Climate change affects the environment, but it also aids in the identification of solutions. In

the past decade, nature has absorbed 54 percent of human-caused CO2 emissions, delaying global warming. Nature can assimilate much more carbon if we stop deforestation, repair ecosystems, manage forests, encourage the soil to store more carbon, and enhance farming techniques. Nature provides protection as well. Healthy ecosystems have the potential to increase humanity's resilience and shield it from the effects of climate change. As with wetlands and mangroves, coral reefs can protect from storm surges. In addition, trees absorb additional precipitation, reducing the risk of discharge, landslides, and flooding-related damage.

We must act swiftly to combat the climate catastrophe. As a consequence of the changes we are currently experiencing, nature is experiencing extensive disruptions. Although some species have adapted to a warmer climate and will continue to do so, other natural systems are at their breaking point. To avoid the most severe effects of climate change, it is essential to limit warming to less than 1.5 degrees Celsius, but every degree of warming counts because the effects increase with each degree. If, for instance, boreal forests encroach upon tundra, fewer of the numerous species that depend on it, such as caribou, arctic foxes, and snowy owls, would have a place to reside. Other changes in the United States include a shift in the boundary between temperate mixed and conifer forests in Southern California, a movement in the boundary between shrubland and conifer forests in New Mexico, and a shift in the boundary between temperate deciduous and conifer forests in Vermont's Green Mountains. As rivers and streams warm, warm water fish are expanding into areas formerly dominated by coldwater species. Warming waters are causing a predicted 47% habitat loss by 2080 for coldwater fish, including numerous coveted trout and salmon species. In certain regions of the western United States, western trout species may lose more than 60% of their entire population, while bull trout populations may lose as much as 90%. Range changes impact the current condition of the ecosystem and may reduce hunting and fishing opportunities (Feng, 2015).

9.7 PHENOLOGICAL CHANGES

Changes in species phenology, or the seasonal timing of life history events, are one of climate change's most visible and least understood effects. We need a theory that can explain current phenological trends and predict future changes as the climate continues to warm and become more unpredictable. We present a mathematical model that links the phenotypic attributes and temperature effects of individual organisms to the population-level phenological patterns observed in ectotherms. It is exhaustive enough to encompass all ectothermic organisms and mechanistic enough to make precise predictions. Its strength lies in its ability to generate projections that are entirely independent of population-level observations of phenological changes and are based solely on temperature-related life history characteristics (Williamson, 2013).

Climate change has one of the most visible effects on phenology. Although a significant number of species exhibit delayed or no change in emergence, it is commonly believed that high-latitude ectotherms have advanced emergence. A mechanistic theoretical framework was presented in a study that reconciles these contradictory data and predicts population-level phenological patterns based solely on how the underlying life history characteristics respond to temperature. The model predicts that optimum abundance will occur earlier in the year if the average ambient temperature increases. Peak abundance occurs later in the year when the amplitude of seasonal oscillations is amplified by warming. When summer temperatures exceed the maximum for growth and reproduction in high-latitude species, we discovered that this does not always result in a prolonged activity period. The results support and refute the following hypothesis regarding ectotherm species affected by climate change. An increase in mean temperature has a greater negative impact on species adapted to high mean temperatures and low-amplitude seasonal fluctuations at low latitudes, whereas an increase in seasonal fluctuations has a greater negative impact on species adapted to low mean temperatures and high-amplitude seasonal fluctuations at high latitudes.

The most influential abiotic factor governing phenology, or the seasonal timing of biological events, is temperature. Increasingly, climate change is interfering with natural phenological cycles, but it is unknown how these disruptions may affect population dynamics and species interactions. Since ectotherms (microbes, plants, invertebrates, fish, amphibians, and reptiles) make up the overwhelming majority of the world's biodiversity, understanding the relationship between their phenology and population dynamics is a critical research priority. Fertility, development, and survival in ectotherms are all physiological responses to temperature fluctuations. As the temperature changes, so does the reactivity. The qualitative forms of these reactions (monotonic, unimodal) are conserved among ectotherm species and are caused by temperature changes in the underlying metabolic processes (such as reaction kinetics, enzyme inactivation, and hormone regulation). To predict how warming-induced phenological shifts will affect the population dynamics of ectotherms, we must focus on these defining responses (Feng, 2015).

The nature of thermal adaptation necessitates that trait responses be concave-up or concave-down functions of temperature. a concave up temperature response function (for instance, exponential) will produce a higher average response in a more variable environment, whereas a concave down temperature response function (for instance, unimodal) will produce a lower average response in a more variable environment. All ectotherms exhibit a unimodal (i.e. concave-down) response to temperature in terms of their per capita birth rates. Higher-latitude species (such as temperate species) with wider response breadths (greater flexibility) perform worse on average than lower-latitude species (such as tropical species) as a result of thermal adaptation to higher-latitude conditions. In this instance, stabilizing selection causes ectotherms at higher latitudes to evolve thermal optimum temperatures that are higher than the average habitat temperature. There are numerous empirical examples of this variation in maximum temperatures (Feng, 2015).

In light of these latitudinal differences in temperature tolerance, the effects of global warming should vary across latitudes. Because the thermal optimum of tropical ectotherms coincides with the mean habitat temperature, the birth rate will be reduced and the mortality rate will be higher for the majority of the year. As a consequence, when the average temperature rises, these optimums decline below the mean. The species will become extinct if the average temperature rises to the point where the mortality rate exceeds the birth rate. A small increase in average temperature is likely beneficial, as the thermal optimum temperatures of temperate ectotherms are higher than the mean temperature of their habitat. Consequently, it is anticipated that tropical ectotherms will be more negatively affected by global warming due to an increase in average temperature than temperate ectotherms. On the other hand, temperate ectotherms are predicted to suffer more from warming that increases seasonal temperature variability [such as an increase in mild and cold extremes] than tropical ectotherms. The average fatality rate must be greater when seasonal fluctuations are greater due to the concave-up character of the mortality response. Moreover, the concave-down character of the birth rate reaction results in a decrease in the average birth rate. If warming causes, the amplitude of variations to exceed the response breadths of a species, the average mortality rate will exceed the average birth rate, putting the species at risk of extinction. As a consequence of the qualitative nature of life cycle characteristic responses to temperature, we can anticipate the diverse effects of global warming on the survival of ectotherm species at various latitudes.

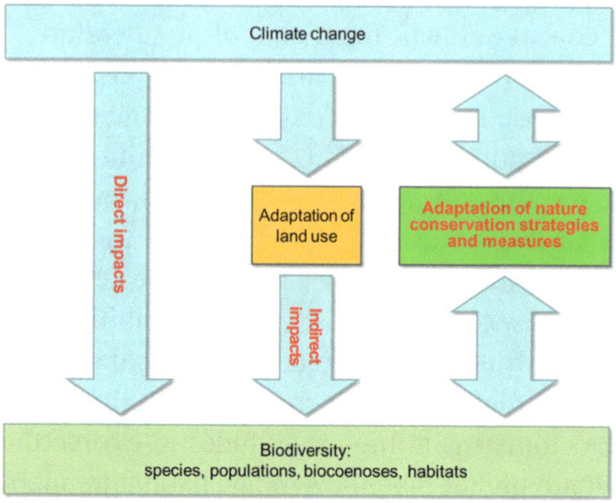

Figure 9.5: Climate-change impacts on biodiversity.

Source: Schliep R, Walz U, Sukopp U, Heiland S. Indicators on the Impacts of Climate Change on Biodiversity in Germany—Data Driven or Meeting Political Needs? *Sustainability*. 2018; 10(11):3959. https://doi.org/10.3390/su10113959 [accessed 12 Jun, 2023]

Predicting the effects of warming on phenology, however, is more challenging because temperature influences on development must be taken into account. Over 80% of metazoan animals have complex life cycles that are influenced by temperature-induced developmental delays. These developmental delays eventually determine the phenological events (such as the timing of emergence and reproduction) observed at the population level (e.g., the timing of emergence and mating). To calculate species' per capita growth rates (fitness), the metric that translates temperature effects on life history traits into population-level outcomes, it is necessary to comprehend how the temperature response of development interacts with the temperature responses of birth and mortality rates. To accomplish this, it is necessary to construct a mathematical framework capable of scaling from trait responses to population dynamics.

Even among species that inhabit the same latitude, phenological fluctuations are a consequence of climate change. Understanding how temperature influences phenological patterns may help to explain these differences (Feng, 2015).

9.8 DISRUPTION OF ECOSYSTEM SERVICES

Temperature increases alter ecosystems by affecting both the timing of the seasons and the geographic distribution of particular habitat categories. Populations of European butterflies, for instance, migrated 114 kilometers north between 1990 and 2008 as a consequence of higher temperatures and the expansion of suitable habitats, according to a study. Lowland species migrate northward as a result of rising temperatures. As temperatures rise, many species can no longer flourish in their former habitats. According to experts, climate change alone threatens 8% of the world's animal species with extinction. Numerous species cannot acclimate to rising temperatures near the equator, which is home to the greatest biodiversity on Earth. As the water continues to warm, reef fish will expire because they are already existing in the hottest water they can endure. In addition, by 2070, temperatures exceeding their upper limit are anticipated to inhibit the germination of nearly 20 percent of tropical plant species (Williamson, 2013).

As the climate increases, scorching, dry conditions will occur more frequently, which will exacerbate wildfires. According to projections, the average area burnt by wildfires in the western United States will increase by 600% for every 1°C increase in global temperature, threatening the local populations of flora and animals in the path of the fire. An estimated one billion animals were destroyed by the conflagration that ravaged 25 million acres in Australia in 2019-2020. During an abnormally scorching and arid season, a bolt of lightning ignited the fire. As many of the species that succumbed to these fires are endemic to Australia, their survival is cause for concern. Although local plant and animal populations typically recover after a fire, species found in extremely small numbers or in specialized locations are more susceptible to global extinction. According to specialists, climate change is responsible for the increase in extreme weather events, which have a swift and significant impact on species and ecosystems. Temperatures of 42°C (approximately 108°F) killed one-third of the spectacled flying foxes in Cairns, Australia, in just two days during a heatwave in 2019. Even though they are accustomed to the constant heat of Australia, animals such as flying foxes cannot tolerate the extreme temperatures that are becoming more frequent due to climate change.

Warmer sea surface temperatures are increasing the intensity of tropical cyclones and hurricanes, thereby modifying or eliminating the habitats of coastal plants and animals. These cyclones can affect ecosystems by interfering with the nitrogen cycle and plant growth due to their intense flooding and winds. As a result of global warming, tropical cyclones, and hurricanes are striking more locations than in the past, exposing more ecosystems to the hazards of extreme weather. Many coastal communities are being evacuated due to rising sea levels, which are also damaging wetland ecosystems. As a consequence of rising sea levels, up to 22 percent of the world's wetland areas may disappear by 2080. Coastal Louisiana, which contains more than two million acres of wetlands, is losing approximately one football field's worth of land every 45 minutes, which is quicker than almost anywhere else. Wetland ecosystems are essential for mitigating climate change because they store three times as much carbon as forests and shield coastlines from flooding. Loss of coastal wetlands as a result of climate change endangers the diverse flora and fauna that inhabit the area and harms fisheries businesses that depend on marine life (Feng, 2015).

Climate changes have also affected important ecosystem-level characteristics and attributes that typically vary spatially. For instance, primary producers, or organisms capable of converting sunlight into energy, are essential to virtually all life on Earth. These animals, which also produce oxygen, constitute the majority of food webs. Warming ocean temperatures in lukewarm waters are limiting the transport of vital deep-ocean nutrients, which are expected to increase phytoplankton's primary production, which accounts for half of the ocean's primary output. In contrast, the melting of sea ice makes more sunlight available, promotes the development of phytoplankton populations, and boosts primary production in climates adjacent to polar seas. Different species' exposure to and responses to climate change are influencing the evolution of interspecies interactions, which has the potential to affect the structure and function of ecosystems. Moreover, variations in the duration, intensity, and frequency of extraordinary events such as droughts, forest fires, and heat waves have had a substantial impact on ecosystems (Williamson, 2013).

Climate change may have an impact on the benefits and services provided by natural ecosystems to humanity. Although not all outcomes are negative, even positive developments may necessitate costly societal adjustments. The effects of changes in primary production on the aquatic food web, for instance, will affect the maintenance of sustainable fisheries. The interaction between rising temperatures and rising nutrient levels exacerbates the severity of harmful algal blooms. In addition, the influence of climate change on the capacity of ecosystems to provide labor opportunities, leisure opportunities, and restorative experiences would negatively impact the physical and mental health of communities. The degree to which changes in ecosystem services impact human welfare will depend in part on the adaptability of human communities to such changes. To reduce long-term regional repercussions, natural resource managers need proactive, adaptive adaptation strategies that consider both past and prospective data, but their use remains unequal across the nation (Feng, 2015).

The impacts of climate change on ecosystem services and biodiversity are severe. However, because various species, populations, and even entire ecosystems react differently to climate change, it can be difficult to devise effective adaptation techniques.

9.9 CLIMATE CHANGE IMPACTING MOUNTAIN HABITATS

Mountains, which constitute over a quarter of the Earth's landmass, serve as a habitat for over one-tenth of the global population. Global climate change presents various potential threats to mountain ecosystems. As per a research analysis, the phenomenon of climate change is anticipated to have a profound impact on both lowland and mountain ecosystems, leading to alterations in the frequency and severity of forest fires, the diversity of fauna, and the distribution of unpolluted water. Based on scientific investigation, elevated temperatures could result in the relocation of habitats situated at lower elevations towards the upper alpine zone. The implementation of such a modification has the potential to disturb infrequent alpine meadows and other elevated ecological systems. To acclimate to persistent alterations in the surrounding ecosystem, the indigenous plant and animal species must undertake a vertical migration towards higher elevations, where the availability of unoccupied habitats is limited. Various groups of species worldwide have undergone both range and population expansions (Williamson, 2013).

The impact of climate change is evidenced by the melting of glaciers, which in turn leads to a reduction in the depth of mountain snowpack. Variations in the thawing patterns during different seasons can potentially exert a noteworthy impact on mountainous areas that are dependent on the discharge of freshwater. The temporal and spatial patterns of discharge could potentially undergo alterations in the event of accelerated and premature snowmelt during the spring season, which can be attributed to the escalating temperatures. The alterations in question have the potential to impact the accessibility of freshwater resources for both human and ecological systems (Feng, 2015).

Climate change is the main cause of forest fires in both flourishing and deteriorating forests. Indonesian forest fires have escalated since 1997. To clear land faster, farmers often ignite fires. These fires produce 15% of fossil fuel combustion's annual CO2 emissions, according to estimates. Carbon dioxide increases tree growth, according to 2018 research. The trees have lost eight to twelve percent of their density since 1900. Despite rising timber output, the authors argue that it contains less material than decades before. Arctic areas are more sensitive and warming faster. Smoke particles on snow and ice can absorb solar energy that would normally be reflected, promoting warming. Fires warm Arctic permafrost, releasing methane. Modernizing forecasting systems will fix the problem (Williamson, 2013).

Figure 9.6: Typical alpine tundra found at Niwot Ridge (ca 3500 m.a.s.l., 40°3′ N Lat., 105°35′ W Long). The dry meadow in the foreground has been snow-free for nearly a month, while snowbed communities and some of the moist meadows have yet to emerge from the snow. Willow (here *Salix glauca* and *S. planifolia*) has increased exponentially at this site in recent decades ([**26**]; photo by C. M. Tate).

Source: Seastedt TR, Oldfather MF. Climate Change, Ecosystem Processes and Biological Diversity Responses in High Elevation Communities. *Climate*. 2021; 9(5):87. https://doi.org/10.3390/cli9050087

9.10 ALTERED OR REDUCED ECOSYSTEM SERVICES

Some of the crucial functions that ecosystems provide for civilization are being impacted by climate change. Ecosystems, for instance, provide an abundance of sustenance to individuals. Drought and heat caused by climate change may impact the ability of producers to cultivate diverse commodities and the availability and quality of specific foods.

Climate changes may also impact ecological services such as carbon capture and storage. By absorbing carbon dioxide from the atmosphere and depositing it in their roots, sediment, and forest floor, forest ecosystems play a significant role in the carbon cycle. However, climate-driven increases in wildfires, inundation, parasites, and diseases may hinder an ecosystem's capacity to perform this essential function.

- *Changes in Populations and Species*: Some species will modify their behavior, morphological composition, or physiological processes in response to environmental changes. Certain individuals will be unable to acclimate. Due to climate change, certain populations may expand, contract, or even become extinct. Consequently, these alterations may affect the regional biodiversity as a whole.
- Plants and animals may alter their geographic distribution in response to a changing climate. Numerous plant and animal ranges have already shifted as a result of fluctuating temperature and water conditions. Certain terrestrial species have moved an average of 3.8 miles per decade to the (typically milder) north as temperatures in the United States have risen. Additionally, certain marine species have moved northward at a rate of approximately 17 kilometers per decade (Williamson, 2013).
- *Variations in the Timing of Natural Events and Cycles*: Various stages of the life cycles of numerous plants and animals are triggered by environmental cues, such as temperature and water content. These indicators may alter at varying rates or not at all when the environment alters. Consequently, species that rely on one another during specific seasons may no longer be in synchrony. Plankton, for instance, is an essential food source for juvenile fish, although it frequently adapts to temperature changes faster than fish. This indicates that plankton may not be as readily available when maturing fish require it the most. Moreover, if a bird migrates at the same time every year, it may find that its primary food source has flowered too early and is no longer accessible due to temperature changes (Feng, 2015).
- *The Interdependencies between Ecosystems Have Changed*: Changes in the climate are also altering how humans and animals

interact with their surroundings and one another. These effects can harm an entire ecosystem.

For instance, climate change accelerates the proliferation of exotic species in certain regions. Non-indigenous species are invasive species. Invasive species are capable of displacing native species, introducing new maladies, and causing additional issues. These modifications may negatively impact the economy and the environment.

As ocean temperatures rise, invasive fish species such as the tropical lionfish are anticipated to migrate north along the Atlantic coast, posing a threat to native species. Because lionfish are poisonous and can attack, they can cause harm. Changes in the climate also affect food webs. A food web is the entire network of nutritional relationships between organisms in an ecosystem. Plants and microorganisms are examples of organisms at the bottom of the food chain. They provide sustenance for species higher up the food chain. Any component of a food web may be affected by climate change, which can have repercussions for the entire system and even other ecosystems.

In conclusion, global climate change might cause extensive biodiversity loss on maritime islands and their distinctive and diversified flora. Therefore, the effects of climate change on islands merit increased scientific attention, not least because oceanic islands are well-suited for a variety of climate change-related scientific approaches and have the potential to move beyond their current role as "test tubes" for ecology and evolution to serve as model systems for basic and applied climate change research. Climate change on oceanic island ecosystems has not been thoroughly investigated till now. To accurately assess climate change's ecological impacts, ecosystem changes must be detected and linked to temperature variations. A reliable collection of frequent time-lapsed measurements is needed to identify environmental variable changes. climatic change affects plants in complicated ways, and a species' presence or disappearance in a region depends on variables other than its climatic tolerance. The temperature often limits a plant's capacity to compete. Climate change may affect plant-pollinator, mycorrhizae-herbivore, and pathogen connections. Rising temperatures are the most obvious consequence of climate change, but precipitation changes and infrequent catastrophic occurrences may be more significant. Plant connections and performance are affected by atmospheric composition, notably carbon dioxide levels. Most plants are moving to milder places as climate change alters their phenology and distribution patterns (Feng, 2015).

Climate warming worsens catastrophic occurrences rather than everyday situations. local anthropogenic disturbances, mean climate trends, and climatic extremes interact synergistically and must be addressed for successful biodiversity conservation management. Dung beetle seed dispersion and parrotfish grazing help tropical forests and coral reefs recover. Preserving or restoring ecological complexity and structure locally helps reduce catastrophic disasters. The ability to preserve, repair, and use ecosystems to mitigate climate change has gained popularity within the wider context. Climate change affects ecosystems' ability to supply "adaptation services."

Researchers analyze the biodiversity-ecosystem carbon sink link geographically. Biodiversity is multidimensional and changeable, making mapping harder than carbon. The Arctic is rich in biodiversity, yet tropical forests have more species. They generate proactive and reactive biodiversity conservation potential maps using a range of metrics. Carbon-focused afforestation of natural grasslands may harm biodiversity. The research focuses on tropical and boreal forest zones, not grasslands, where biodiversity and carbon objectives overlap. Griscom et al. explored the national potential for natural climate solutions in tropical nations, where trees provide a large carbon sink and natural climate solutions has the largest climate change mitigation potential. They consider protecting forests, peatlands, mangroves, and working fields. Twelve natural climate solutions pathways might mitigate climate change, biodiversity, and other ecosystem services by limiting forest conversion. Tropical natural climate solutions potential is concentrated in a few nations, all of which have above-average governance measures, making protect-manage-restore techniques feasible. Studies of the practicality of ecosystem-based urban climate change adaptation done. Cities will house two-thirds of the world's population by 2050, making them key climate change hubs. Cityscapes are vulnerable to climate change due to their low vegetative cover, high imperviousness, pollution production, heat island effects, high demand for freshwater, and concentration of infrastructure and population in vulnerable areas like coastal zones, river floodplains, and deforested hillsides. Climate change hazards and urban amplification can be reduced. These methods include building stormwater basins, bioswales, green roofs, riparian zones, and plant cover and green space (Feng, 2015).

Humans have affected big herbivores and predator groups for 50,000 years. Domestic ruminant grazing animals have replaced non-ruminant herbivores in many locations, changing plant structure, fire patterns, and biogeochemical cycles, notably the carbon cycle. Rewilding extant native herbivores to replace ruminant livestock would cut methane emissions, a strong greenhouse gas, but whether they would mitigate climate change depends on the biome. Living native herbivore situations are unlikely to boost natural climate solutions but can benefit ecology and biodiversity. Researchers also underline that rewilding for ecological complexity and biodiversity does not give advantages. Introducing grazing and browsing megafauna (horses, bison, and cattle) to Arctic high boreal and tundra ecosystems is a unique, rare, and dramatic climate change mitigation method. Such an introduction may help restore the "mammoth steppe," a huge high-altitude grassland biome thought to have been lost with the demise of high-latitude Pleistocene megafauna, which human hunting civilizations are thought to have caused. By delaying permafrost collapse and carbon and methane emissions, these high-latitude grassland ecosystems may lower the probability of severe climate change positive feedback in Arctic permafrost. The authors add that Arctic "land use" options may have the same climatic influence as land use at mid- and low-latitudes, which has been thoroughly investigated. Like other natural climate solutions, it is difficult to apply on a big enough scale to affect climate (Williamson,2013).

Chapter

10 PRINCIPLES OF HAZARD MANAGEMENT

CONTENTS

10.1: INTRODUCTION

Hazards are things that could hurt you or put you at risk and have the potential to have negative impacts. They cover a wide spectrum of occurrences, phenomena, or circumstances that put the safety of people, property, and the environment at risk. Hazards can be caused by biological agents, technical processes, environmental variables, or natural forces, and they can have a wide range of effects.

To protect people, property, and the environment, the vital discipline of hazard management focuses on identifying, evaluating, and minimizing potential risks and threats. The framework for creating comprehensive strategies and frameworks to foresee, prevent, prepare for, respond to, and recover from various dangers is provided by hazard management principles. These suggestions assist people, organizations, and communities in navigating the challenges of hazard management and minimizing the negative effects of events like accidents and natural disasters (Ceacero-Moreno, 2021).

Hazards might be natural, technical, environmental, biological, etc. in the modern world. Events like earthquakes, storms, floods, wildfires, and tsunamis are examples of natural hazards. Chemical spills, nuclear accidents, industrial accidents, and transportation mishaps are all examples of technological hazards. Environmental risks include those brought on by pollution, deforestation, climate change, and habitat damage. Epidemics, pandemics, and the spread of contagious illnesses are examples of biological risks.

To address these many risks, hazard management concepts offer a methodical and organized strategy. They include a number of essential elements that improve the efficiency of hazard control techniques. These elements consist of assessing and analyzing risks, preparing and getting ready, taking preventative and mitigating action, and responding and recovering. The first stage in hazard management is risk assessment and analysis. It entails detecting and assessing potential risks, comprehending their possibility and potential repercussions, and evaluating the exposure and susceptibility of populations and infrastructure. Stakeholders can efficiently prioritize their efforts, allocate resources, and make decisions to minimize and manage hazards by performing a thorough risk assessment (White, 2012).

Planning and readiness are essential components of hazard management. This principle emphasizes the significance of creating communication strategies, evacuation plans, business continuity plans, and emergency response plans. To make sure that people are well-equipped to respond to emergencies, comprehend their roles and duties, and effectively communicate and coordinate during crisis circumstances, adequate training and education are crucial.

Through mitigation and preventative techniques, hazards and their potential impacts are meant to be diminished or eliminated. This concept includes the use of engineering controls, administrative controls, and personal protective equipment. Environmental preservation and land use planning are also crucial for the development of resilient communities and the reduction of hazard risk. Response and recovery efforts aim to control and mitigate hazards' consequences as they materialize. This strategy includes, for instance, the establishment of incident command systems, emergency operations centers, search and rescue operations, medical response and care, and psychological support. It is crucial to build a coordinated and effective response to emergencies in order to aid the recovery process for those who have been impacted and their communities (Ceacero-Moreno, 2021).

Collaboration and involvement from several sectors are necessary for hazard management principles implementation. Governments are essential in creating the legal and regulatory frameworks, policies, and procedures that aid in hazard management initiatives. Raising awareness, performing risk assessments, and fostering resilience at the local level all require active community involvement. Through corporate social responsibility programs, public-private collaborations, and business continuity planning, the private sector also contributes significantly.

The ability to handle hazards has been significantly improved by technological breakthroughs. Early warning systems give people and communities timely alerts and information, enabling them to take preventative action. Big data analytics, remote sensing, and geographic information systems all help decision-makers better comprehend and visualize hazards. Technologies utilizing artificial intelligence and machine learning present prospects for developing more precise risk assessments, prediction models, and reaction plans (White, 2012).

The hazard management principles offer a framework for successfully managing hazards and safeguarding people, property, and the environment. Stakeholders can build effective strategies, improve readiness, put mitigation measures into place, and effectively handle emergencies by adhering to these principles. To ensure the safety and resilience of societies in the face of numerous risks, hazard management is a dynamic and continuing process that necessitates constant review, adaptation, and development.

Hazards pose major threats to persons, property, and the environment in our intricately interconnected world. One of the most important foundational components of efficient hazard management is comprehending the nature of dangers and their possible effects. The purpose of this introduction is to give a general overview of risks, including their various categories and the guiding principles of hazard management.

Understanding and analyzing hazards and their aspects, prospective effects, and the vulnerabilities of the impacted communities and infrastructure are essential for managing risks successfully. This knowledge serves as the foundation for creating strategies to reduce risks, improve readiness, respond quickly and effectively, and promote healing and resilience-building. Hazards come in many different forms and have many different aspects, including natural, technical, environmental, and biological concerns. Recognizing risks' presence, traits, and potential effects is necessary for understanding them. Understanding the various hazards will help stakeholders create efficient hazard management plans that will reduce risks, save lives, and create resilient communities (Ceacero-Moreno, 2021).

Numerous dangers pose concerns to infrastructure, environmental protection, and human safety in our dynamic and interconnected world. Effective hazard management and the creation of appropriate mitigation measures depend on an understanding of the various categories of hazards. An overview of the main categories of hazards, including natural, technical, environmental, and biological dangers, is given in this introduction.

10.2 NATURAL HAZARDS AND THEIR MANAGEMENT STRATEGIES

Natural hazards are incidents or processes that take place in the environment naturally and are unavoidable. Usually, geological, climatic, or geophysical processes are to blame. Natural disasters can have catastrophic effects on ecosystems and communities, causing deaths, property damage, and environmental degradation.

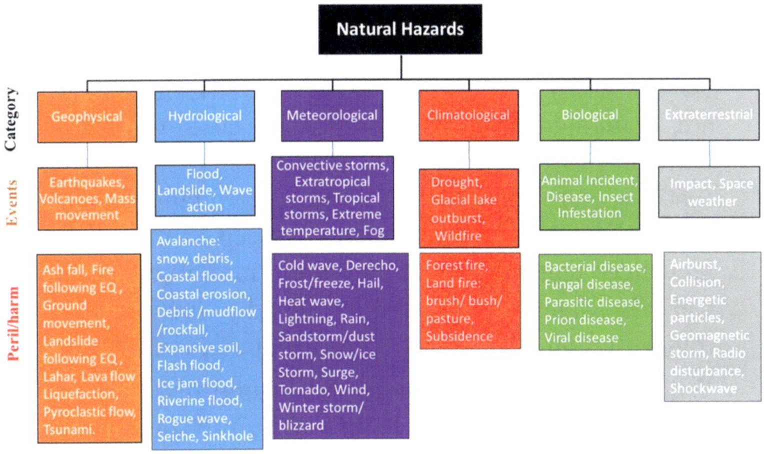

Figure 10.1: Grouping of natural hazards.

Source: Chaudhary MT, Piracha A. Natural Disasters—Origins, Impacts, Management. *Encyclopedia*. 2021; 1(4):1101-1131. https://doi.org/10.3390/encyclopedia1040084 [accessed 9 Jun, 2023]

The following are a few typical natural dangers:

- Geophysical risks: Geophysical risks are caused by internal processes of the Earth and have the potential to drastically alter their surroundings. One of the most well-known geophysical dangers, earthquakes, are caused by the abrupt release of energy in the Earth's crust. Severe earthquakes can also cause tsunamis and landslides. Another geophysical risk brought on by the migration of magma from the Earth's mantle to the surface is volcanic eruptions. Ashes, gases, and pyroclastic flows from these eruptions have the potential to harm adjacent towns and hinder aviation transport (White, 2012).
- Climate-related risks include weather patterns and atmospheric conditions, which are frequently impacted by climate change. Strong climatic risks like hurricanes, cyclones, typhoons, and severe storms are characterized by high winds, a lot of rain, and storm surges. These severe weather incidents may result in significant infrastructure damage, flooding, and fatalities. Another climate risk is heat waves, which are defined by protracted periods of intense heat and can have serious health repercussions, especially for vulnerable populations.

- Hydrological Hazards: Hydrological hazards can happen as a result of excessive rainfall, snowmelt, or dam failures. They are caused by water-related processes. Riverine and flash floods are both significant hydrological hazards that can seriously harm infrastructure and property, interfere with transportation systems, and even result in fatalities. Dam failures can also cause devastating floods downstream. In steep or mountainous areas, landslides that are caused by intense rainfall or seismic activity can also happen, endangering nearby towns and important transportation networks (Ceacero-Moreno, 2021).
- Wildfires: Wildfires are potential natural disasters that may be sparked by lightning or human activity. They quickly spread through shrublands, grasslands, and woods as a result of the wind and dry vegetation. The destruction of habitats, loss of biodiversity, and damage to buildings and infrastructure are just a few of the negative effects that wildfires can have on the environment and the economy. The air quality can be negatively impacted by wildfire smoke, endangering the health of adjacent communities.

Natural hazard management and mitigation call for a multifaceted strategy. This comprises:

- Identification of places vulnerable to particular natural hazards using scientific analysis and mapping techniques.
- Early warning systems are being developed to identify and notify communities about approaching natural disasters, giving people enough time to flee or take preventative action.
- Implementing rules to restrict growth in risky regions and lowering exposure to potential risks is known as land use planning and zoning.
- Engineering solutions: Building resilient infrastructure and structures, such as flood-control systems or earthquake-resistant plans, to endure natural disasters.
- Ecosystem management is the preservation and restoration of natural ecosystems like wetlands and forests, which can act as a natural defense against certain risks like erosion control or flood reduction.
- Education and information: Spreading the word about

precautionary measures and developing a climate of safety and resilience in local communities.

We can lessen susceptibility and lessen the effects of catastrophic occurrences on human lives and the environment by comprehending the nature of natural hazards, putting good hazard management measures into practice, and encouraging community resilience (White, 2012).

10.3 MANAGEMENT OF TECHNOLOGY-RELATED HAZARDS

These risks are connected to infrastructure, activities, and procedures that were created by humans. They result from the use of technology, the handling of dangerous substances, or equipment and system malfunctions. Technological risks can result in mishaps, explosions, releases of dangerous materials, or infrastructure failures, harming people, property, and the environment. Some dangers posed by technology include:

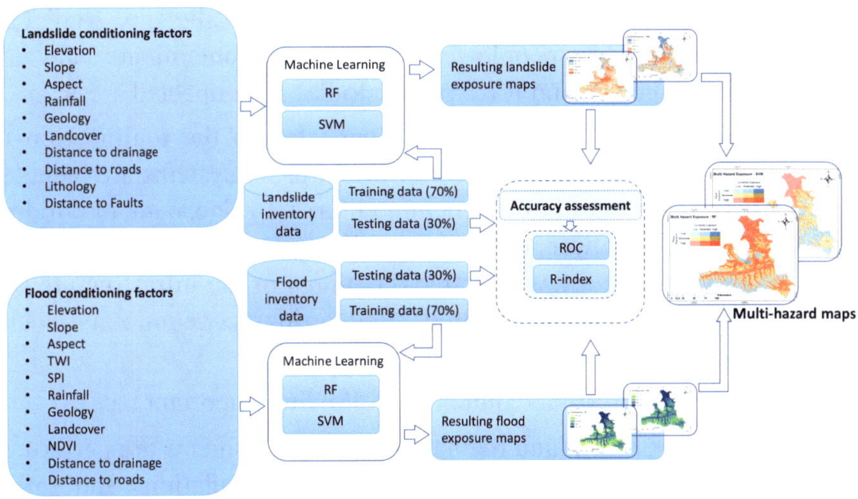

Figure 10.2: Framework for mapping the multi-hazard risk of coupling of natural and technological hazards.

Source: Nachappa TG, Ghorbanzadeh O, Gholamnia K, Blaschke T. Multi-Hazard Exposure Mapping Using Machine Learning for the State of Salzburg, Austria. *Remote Sensing*. 2020; 12(17):2757. https://doi.org/10.3390/rs12172757 [accessed 9 Jun, 2023]

- Industrial Mishaps: Industrial mishaps can happen in factories, chemical plants, refineries, and other industrial settings. Chemical spills, explosions, fires, and defective equipment can all be factors in these mishaps. Environmental contamination, infrastructure damage, and health concerns to neighboring people are just a few of the negative effects that can result from the discharge of dangerous substances into the environment (Ceacero-Moreno, 2021).

- Transport-related incidents: These include dangers connected with the movement of commodities, persons, and hazardous materials. Accidents involving planes, ships, trains, or vehicles transporting hazardous materials can result in fires, explosions, chemical spills, or the discharge of harmful substances. These occurrences might be on land, at sea, or in the air, posing risks to property, the environment, and human life.
- Nuclear accidents: Radioactive material is released in nuclear accidents from nuclear power plants or storage sites for nuclear waste. These can be brought on by defective machinery, careless operators, severe weather, or intentional sabotage. Radiation leaks can harm people for a long time, contaminate land and water supplies, and force people to flee the impacted areas.
- Infrastructure failures: Risks connected to the malfunction or breakdown of vital networks, structures, or systems are included in the category of infrastructure failures. Bridge, dam, or building collapses, power outages, or disruptions in communication networks are a few examples. Failures in the infrastructure can result in accidents, fatalities, and serious economic and social consequences (White, 2012).

Managing technical risks requires the following important tactics:

- Risk assessment and regulation: Risk assessments are performed to detect potential dangers, and strict regulations and safety standards are implemented to reduce risks in the transportation and industrial sectors.
- Plans for emergency response and safety procedures: To reduce the impact of technical dangers and ensure a quick and coordinated reaction during emergencies, detailed safety protocols, contingency plans, and emergency response processes should be established.
- Education and training: delivering the proper instruction and training regarding safety precautions, hazard awareness, and

response procedures to employees, emergency responders, and the general public (Ceacero-Moreno, 2021).

- Monitoring and inspections: To identify possible risks, maintain compliance with safety requirements, and proactively address hazards, routine monitoring and inspections are conducted on industrial facilities, transportation systems, and vital infrastructure.
- Adopting technological innovations to improve hazard detection, early warning systems, and emergency response capabilities, such as remote monitoring, sensor networks, and predictive analytics.
- Public-private partnerships: Encouraging cooperation between governmental institutions, commercial enterprises, and neighborhood associations to share knowledge, resources, and experience in addressing technological risks.

Technology risks can reduce the likelihood of accidents, protect lives, protect the environment, and maintain the resilience of vital infrastructure systems by being proactively identified, assessed, and managed by society.

10.4 MANAGEMENT OF ENVIRONMENTAL RELATED HAZARDS

Risks associated with the degradation, alteration, or pollution of the environment are referred to as environmental risks. These risks are frequently caused by human activity and its effects on ecosystems, the climate, and natural resources. Long-term effects on human health, biodiversity, and the sustainability of the planet can result from environmental risks. Environmental risks include, for example:

- Climate Change: The accumulating greenhouse gases in the atmosphere, primarily brought on by human activities like burning fossil fuels and deforestation, pose a significant environmental risk. Temperatures rise as a result, and precipitation patterns, sea level, and the frequency and intensity of extreme weather events all change. Climate change poses a threat to vulnerable populations, ecosystems, agriculture, water supplies, and human health.
- When dangerous compounds contaminate the air, water, or soil, pollution, a pervasive environmental threat, results. It can come from various sources, including industrial processes,

vehicle emissions, inappropriate waste disposal, agricultural operations, and more. Common types of pollution include soil contamination, water contamination from chemicals, heavy metals, and microplastics, and air pollution (including smog and particle matter). The functioning of ecosystems, biodiversity, and human health can all be adversely affected by pollution (White, 2012).

- Deforestation is the term used to describe the clearing and removal of forests, typically for logging, urbanization, and agricultural expansion. Due to this environmental risk, habitats are lost, soil is eroded, biodiversity is decreased, and greenhouse gas emissions are raised. In addition to disrupting local and global ecosystems, deforestation can alter the availability of resources, water cycles, and climate regulation.
- Habitat Degradation: Habitat degradation is the degradation or destruction of natural ecosystems such as wetlands, coral reefs, and forests. Human activities, such as urbanization, infrastructure construction, mining, and unsustainable agriculture methods, frequently cause this hazard. Ecosystem disruption, biodiversity loss, and compromised ability to provide services like water filtering, carbon sequestration, and support for pollinators are all consequences of habitat degradation (Ceacero-Moreno, 2021).

Managing environmental risks requires a variety of tactics:

- Implementing strategies to preserve natural habitats, safeguard threatened species, and rebuild damaged ecosystems.
- Promote sustainable methods in forestry, agriculture, fisheries, and other resource-intensive businesses to lessen their negative effects on the environment.
- Implementing policies, procedures, and methods to reduce pollution emissions and support cleaner manufacturing processes is known as pollution prevention and control.
- Transitioning to renewable energy sources and reducing reliance on fossil fuels can help to reduce air pollution and combat climate change.
- Assessment of the environmental impact of planned projects and policies in order to spot and address any potential environmental risks.

- Raising public understanding of environmental risks, encouraging sustainable lifestyles, and encouraging a sense of duty for environmental stewardship are all part of public awareness and education.
- Cooperation at the international level to handle transboundary environmental dangers, exchange expertise, and create coordinated environmental protection policies.

Societies can reduce risks, safeguard ecosystems, and guarantee the long-term health and well-being of the planet and its inhabitants by identifying the effects of environmental dangers and putting sustainable policies into place (White, 2012).

10.5 STRATEGIES FOR BIOLOGICAL HAZARDS MANAGEMENT

Risks connected to biological agents, such as viruses, bacteria, parasites, and poisons, are referred to as biological hazards. These dangers pose serious risks to the general public's health and social order by contributing to the spread of infectious illnesses, epidemics, or pandemics. Biological dangers include, for instance:

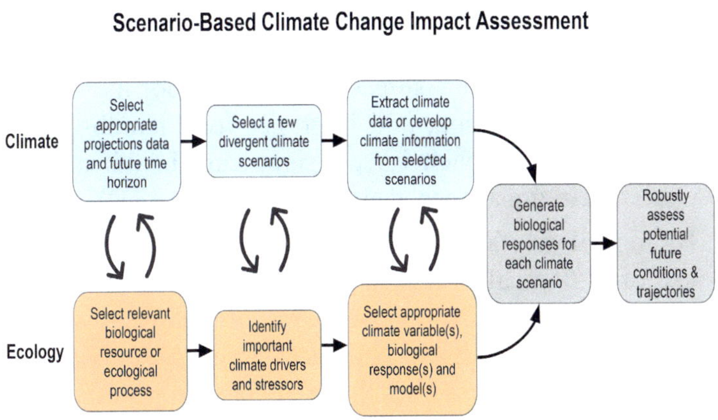

Figure 10.3: Process flow diagram of a typical approach for conducting biological impact assessments under different future climate scenarios.

Source: Rangwala I, Moss W, Wolken J, Rondeau R, Newlon K, Guinotte J, Travis WR. Uncertainty, Complexity and Constraints: How Do We Robustly Assess Biological Responses under a Rapidly Changing Climate? *Climate*. 2021; 9(12):177. https://doi.org/10.3390/cli9120177 [accessed 9 Jun, 2023]

Pandemics and epidemics: Pandemics and epidemics are serious biological risks marked by the swift spread of infectious illnesses. While pandemics happen when an epidemic spreads across multiple countries or continents, affecting a significant amount of the world's population, epidemics relate to the breakout of a disease that affects a particular region or community. Examples include the COVID-19 pandemic brought on by the SARS-CoV-2 virus as well as the 1918 influenza pandemic (Spanish flu), the 2009 H1N1 influenza pandemic, and the current COVID-19 pandemic. Both epidemics and pandemics have the potential to cause significant morbidity, death, social unrest, and economic effects (Ceacero-Moreno, 2021).

- Diseases that are transmitted to people through the bite of infected vectors, such as mosquitoes, ticks, fleas, or flies, are known as vector-borne diseases. Malaria, dengue fever, Zika virus, Lyme disease, and the West Nile virus are some of these illnesses. The distribution and prevalence of vectors and the diseases they transmit are impacted by a number of variables, including climate change, urbanization, and globalization.
- Foodborne Illnesses: Consuming contaminated food or water exposes a person to a biological risk known as a foodborne sickness. Salmonella, *E. coli*, norovirus, hepatitis A, parasites like Toxoplasma and Cryptosporidium, or toxins made by microbes like botulinum toxin can all be responsible for these risks. Foodborne infections spread because of poor hygiene practices, inadequate sanitation, and improper food handling. Such risks can result in stomach issues, significant health issues, and, in extreme circumstances, death.
- Biological warfare and terrorism both include the intentional use of biological agents to injure, frighten, and disrupt people. These dangers might include the use of biological weapons or the deliberate dispersal of infectious diseases. Anthrax, smallpox, the ricin toxin, or viruses that have undergone genetic modification are a few examples of potential biological threats. Due to their purposeful nature and the requirement for improved surveillance, readiness, and reaction capabilities, biological warfare and terrorism present special difficulties (White, 2012).

Managing and reducing biological threats requires the following important tactics:

- Establishing reliable surveillance systems to find and track disease outbreaks will enable early detection and action. Disease surveillance and monitoring.

- Health promotion actions putting in place public health measures to stop the spread of infectious diseases, such as vaccination programs, disease testing, contact tracing, and quarantine procedures.
- Implementing strategies such as insecticide use, habitat management, and mosquito control programs in order to stop the spread of vector-borne diseases.
- Sanitation and hygiene procedures: encouraging appropriate sanitation, safe food handling, and good hygiene habits to stop the spread of infectious diseases and foodborne illnesses.
- In order to prevent unintentional releases of and unauthorized access to potentially harmful biological agents, laboratories, healthcare facilities, and research institutes must implement the necessary safety measures and processes for biosafety and biosecurity.
- Creating thorough plans and strategies for epidemic and pandemic preparedness, including the readiness of the healthcare system, the storing of medical supplies, and coordination between pertinent agencies and stakeholders.
- Research and development: Promoting the creation of vaccinations, therapies, and diagnostic tools as well as tools to improve preparedness and response.

Societies can lessen the effect of infectious diseases, safeguard public health, and improve overall resilience by proactively addressing biological hazards through a combination of prevention, surveillance, readiness, and reaction (Ceacero-Moreno, 2021).

It is essential to comprehend the various threats in order to adopt effective hazard management techniques. Stakeholders can create efficient mitigation, readiness, response, and recovery plans to protect people, property, and the environment by being aware of the unique traits and potential effects of each type of hazard.

10.6 PRINCIPLES OF HAZARD MANAGEMENT

Hazard management is crucial for safeguarding infrastructure, protecting people, and minimizing the effects of potential hazards and threats. Only a handful of the numerous potential causes of hazards include natural disasters, technological failures, environmental risks, and biological concerns. The principles of hazard management provide a framework for understanding, assessing, and dealing with these hazards in a way that lessens their negative impacts. By putting these ideas into practice, people, communities, organizations, and governments can increase their resilience and ability to handle threats. This article examines the fundamentals of risk management and their significance for ensuring safety and long-term development (White, 2012).

10.6.1 Risk Assessment and Analysis

Conducting thorough risk assessments and analyses is the basic concept of hazard management. possible hazards must be identified, their likelihood and possible effects must be assessed, and the exposed and vulnerable systems and populations must be understood. In order to reduce the impact of risks, it is important to prioritize resources, create efficient mitigation plans, and make well-informed judgments.

While risk analysis focuses on quantifying and defining the type and extent of those hazards, risk assessment involves the detection, analysis, and evaluation of risks. Let's explore these ideas in greater depth:

- Identification of Hazards: The identification of hazards is the first phase in the risk assessment process. Recognizing potential sources of harm, such as natural disasters (such as earthquakes, floods, and hurricanes), technological errors (such as chemical spills and infrastructure collapses), environmental threats (such as pollution and deforestation), or biological dangers (such as infectious diseases and epidemics), is a step in this process. A thorough grasp of the spectrum of hazards that need to be assessed is made possible by thorough identification.
- Analysis of Vulnerabilities and Exposures: Following the identification of hazards, the focus of the analysis moves to evaluate vulnerabilities and exposures. To establish their vulnerability and possible exposure to the listed hazards, the elements at risk, such as human populations, infrastructure, ecosystems, and economic assets, are examined in this step. Prioritizing regions or systems that need more care and protection depends on understanding vulnerabilities and exposures (Ceacero-Moreno, 2021).
- A key component of risk analysis is determining the likelihood

that a hazard event will occur. In order to determine the likelihood of a hazard occurrence occurring within a specific timeframe, this entails taking into account variables including historical data, scientific models, and expert judgment. A likelihood assessment gives stakeholders knowledge of the frequency or recurrence of hazards, allowing them to invest resources wisely and put in place the right preventive measures.

- Consequence Assessment: Conducting a consequence assessment entails assessing the severity and probable effects of a hazard event. It takes into account both direct and indirect consequences on social well-being, infrastructure, the environment, and the economy. The effects could result in fatalities, property damage, environmental harm, economic losses, service disruptions, and social unrest. Prioritizing risk reduction activities and developing effective reaction plans are made easier with the use of consequences analysis.
- Risk assessment: To ascertain the degree of risk attached to a hazard, risk assessment blends the likelihood and consequence evaluations. In this stage, the projected risk level is compared to established risk thresholds or criteria. Risk assessment enables decision-makers to rank risks according to their seriousness, likelihood, and potential effects, aiding in resource allocation for risk management efforts.
- Effective risk communication is crucial for effective risk analysis and assessment. Clear and open communication of risk information is required with all relevant parties, including the general public, decision-makers, and relevant organizations. Increased knowledge, awareness, and the capacity to make educated decisions regarding risks and the efforts being taken to decrease them are the main objectives of risk communication. It is feasible to improve trust, facilitate readiness, and encourage proactive measures through prompt and transparent communication (White, 2012).
- Iterative Process and Review: Risk assessment and analysis are continual, iterative processes, not one-time operations. The risk assessment needs to be reviewed and updated as new data, information, or circumstances surface. Risk management techniques are regularly reviewed to make sure they are still

effective and relevant in light of evolving hazard characteristics, vulnerabilities, and exposure patterns. The enhancement of risk assessment procedures is facilitated by ongoing improvement and experience-based learning.

Hazard management practitioners get important insights into the nature and scope of risks connected with hazards by undertaking rigorous risk assessment and analysis. The ability to make well-informed decisions, implement focused risk reduction strategies, and allocate resources efficiently is made possible by this information, which ultimately improves overall resilience and readiness to deal with future risks (Ceacero-Moreno, 2021).

10.6.2 Prevention and Mitigation

Preventing or eliminating the risks brought on by hazards is the goal of prevention and mitigation. This principle places a strong emphasis on taking preventative actions to avoid or reduce the likelihood of risks, such as putting construction norms and regulations into place, creating early warning systems, enforcing safety regulations, and encouraging sustainable land-use planning. The intensity and frequency of potential disasters can be greatly decreased by addressing the underlying causes of hazards and putting preventive measures in place (White, 2012).

Let's explore prevention and mitigation tactics in more detail:

- Implementing and upholding building codes and regulations is a crucial preventive step. These codes specify requirements for the planning, building, and upkeep of infrastructure to withstand potential dangers. They cover topics including electrical and mechanical systems, structural integrity, fire safety, seismic resistance, and flood-resistant design. Building codes assist in ensuring that new constructions are risk-resistant and that older buildings are renovated to increase their resilience.
- Early warning systems: Early warning systems are essential for preventing hazards. These systems use a variety of technology, including sensors, satellites, and meteorological data, to identify emerging dangers and send out prompt alerts to vulnerable people. Early warnings enable people and communities to take preventative measures including protecting property, fleeing risky regions, and finding shelter. Effective early warning systems reduce the severity of risks and save lives.

- Planning for land use entails identifying and allocating regions for particular uses based on how vulnerable they are to threats. It seeks to direct development in a way that reduces risk exposure. Construction may not be allowed in high-risk regions near landslides or flooding, for instance, due to zoning rules. Land-use planning lessens vulnerability and prevents the emergence of new risk-prone areas by diverting growth away from dangerous locations or enacting protective measures.
- Ecosystem management and restoration: Protecting and regenerating natural ecosystems is a successful mitigation tactic. Healthy ecosystems offer built-in defense against dangers, including forests, wetlands, and coastal habitats. They stabilize slopes, act as a buffer against storm surges, absorb and slow flooding, and control climate patterns. Enhancing ecosystem resilience and lowering the susceptibility of neighboring communities to risks are two benefits of ecosystem protection and restoration (Ceacero-Moreno, M. 2021).
- Infrastructure resilience: Improving the critical infrastructure's resilience is essential for hazard reduction. In order to do this, infrastructure such as highways, bridges, electricity grids, and water delivery systems must be designed and built to withstand possible dangers. Structures can be strengthened, hazardous materials can be used, redundant systems can be set up, and adaptable design concepts can be used as measures. Infrastructure that is resilient limits damage, ensures the continuation of important services, and speeds up recovery.
- Raising public awareness and spreading education about risks and how to mitigate them are essential. People and communities are better equipped to take preventative measures and make wise decisions when they are taught about potential dangers, safety precautions, and emergency response protocols. Campaigns for education, training, and community exercises all help to improve readiness, build a culture of safety, and instill a feeling of accountability in hazard management.
- Hazard prevention and mitigation are aided by effective management of the environmental resources. Sustainable techniques assist preserve ecosystem health and lessen the chance of risks like wildfires or soil erosion. These activities

include responsible forestry, sound land management, and soil conservation. Flooding risk and other water-related dangers are reduced by responsible management of water resources, which includes floodplain management and watershed preservation.

- Technological Innovations: New prospects for hazard mitigation and prevention are presented by technological advancements. For instance, hazard-prone locations can be found and decision-making is aided by the use of remote sensing, geospatial analysis, and modeling technologies. Levees, flood barriers, and slope stabilization methods are examples of cutting-edge engineering solutions that improve protection against certain risks. Early warning systems, monitoring networks, and information sharing are all supported by technological developments (White, M. P. 2012).

To lessen the effects of hazards and increase resilience in the face of potential dangers, prevention and mitigation techniques are essential. Individuals, communities, organizations, and governments may actively decrease vulnerabilities, safeguard people and property, and advance sustainable development by putting these measures into place (Ceacero-Moreno, 2021).

10.6.3 Preparedness

A crucial component is preparedness, which entails creating the resources, strategies, and systems needed to respond to dangers. This involves creating emergency response plans, setting up communication systems, doing drills and simulations, and giving communities and individuals training and instruction. When dangers arise, preparedness makes sure that all relevant parties have the information, abilities, and resources they need to react quickly and efficiently, minimizing the loss of life and property.

Let's examine some important components of preparedness:

- Planning an emergency response entails creating detailed plans and procedures to direct actions before, during, and after a hazard event. The roles and responsibilities are outlined, the communication channels are established, the evacuation routes are identified, the emergency shelters are designated, and the resource and personnel deployment is coordinated. Effective and coordinated responses in urgent situations are guaranteed by well-designed response plans.

- Systems of communication and information: Establishing reliable systems of communication and information is crucial for readiness. Providing the public with timely and accurate informationrequires establishing dependable communication networks, such as emergency hotlines, public address systems, and broadcasting channels. Information systems should make it possible for responding agencies to share data effectively, enhancing coordination and decision-making in times of crisis.
- Training and education: It is essential for preparation to offer training and education to individuals, communities, and response personnel. Training programs give people the knowledge and abilities needed to react correctly to risky situations. This includes instruction in first aid and medical procedures, search and rescue methods, fire safety regulations, and emergency preparedness measures. A more efficient and well-coordinated emergency response is facilitated by responding professionals with adequate training.
- Public awareness and community engagement: Two important aspects of preparedness are raising public awareness and encouraging community involvement. People are more equipped to take preventative measures to safeguard themselves and their communities when they are informed about potential risks, their effects, and the required preparedness measures. Participating in hazard management decision-making processes with local populations promotes a sense of ownership, boosts resilience, and promotes teamwork in times of crisis.
- Resource stockpiling and logistics: Being prepared entails strategically storing the goods and resources needed for disaster response. These could contain supplies for an emergency shelter, food, water, and medicine. The prompt deployment and distribution of resources to affected areas, supporting emergency response activities, and meeting the needs of affected populations are all made possible by adequate logistical planning (Ceacero-Moreno, 2021).
- Drills and simulation Exercises: Regular drills and simulation exercises are essential for preparation. Through these simulations of dangerous situations, response organizations, emergency personnel, and communities may hone their response strategies

and tactics. Through simulations, holes are found, communication systems are tested, coordination methods are assessed, and stakeholder readiness is generally improved to handle actual hazard scenarios.

- Collaboration and coordination are essential to being prepared, as they bring together all the parties involved in hazard management. This comprises governmental institutions, emergency services, non-governmental organizations, businesses, and neighborhood-based groups. The efficiency and efficacy of response operations are increased by establishing open lines of communication, collaborative frameworks, and coordinating procedures.
- Planning for continuity: Planning for continuity focuses on keeping vital services and operations running both during and after a risk occurrence. In order to assure their continuous operation or prompt restoration, this entails identifying vital infrastructure, such as electricity generation facilities, hospitals, and transportation systems, and preparing contingency plans. Planning for continuity reduces disruptions and speeds up the recovery process.

Communities and organizations can considerably lessen the effects of hazards by investing in preparedness measures. Being well-prepared improves one's capacity to intervene quickly, save lives, and reduce environmental and infrastructure damage. It encourages adaptability, cultivates a culture of safety, and makes sure of a more well-organized and effective reaction to risk incidents (White, 2012).

10.6.4 Response and Emergency Management

The notion of "response and emergency management" is concerned with the quick decisions made during and after a hazard event. As part of this, emergency services must be called in, rescue and relief operations must be planned, medical aid must be given, impacted populations must be evacuated, and vital infrastructure must be rebuilt. A well-coordinated response aids in reducing the danger's effects, ensuring the security and wellbeing of those who are impacted, and speeding up the healing process.

Let's look at some important components of emergency management and response:

- Emergency coordination and command: Timely and well-organized responses to hazards require efficient coordination and command structures. Establishing emergency operation centers (EOCs) allows for information sharing and collaboration among important decision-makers and response organizations. In order to ensure a coordinated response effort, the EOC acts as a central point for organizing resources, communication, and decision-making.

- Emergency alert systems: In the case of a hazard, the public's safety depends on the prompt transmission of warnings and alarms. To alert the public to an impending threat, evacuation orders, and safety advice, emergency alert systems use sirens, text messaging, mobile applications, and broadcast systems. People can make educated judgments and take the necessary precautions to protect themselves when there is timely and clear communication.
- Operations for Search and Rescue (SAR): SAR operations are a crucial part of emergency response. To find and rescue people in trouble or who are stuck in dangerous situations, trained personnel such as firefighters, paramedics, and specialized rescue teams are sent out. To ensure the safe and effective recovery of survivors, SAR operations require specialized equipment like search dogs, thermal imaging cameras, and extraction tools (Ceacero-Moreno, 2021).
- Medical treatment: Providing medical treatment and support is a crucial component of emergency management. The injured are given access to medical teams that include paramedics, nurses, and doctors right away. This could involve triage, first aid, urgent operations, and patient transfer to hospitals or other healthcare institutions. Coordination of the distribution of drugs, vaccinations, and medical supplies is another aspect of medical assistance.
- Operational procedures for evacuation and sheltering are started when risks represent a direct threat to people's safety. Plans for evacuating impacted populations to designated shelters or safer locations specify safe routes and assembly locations. In order to give persons displaced by the danger occurrence a safe and secure environment, temporary shelters are set up. The safety and protection of the elderly, children, and those with disabilities are given top priority during evacuation and sheltering operations.

- Damage assessment and situation analysis: To comprehend the scope of the impact and focus response efforts, damage assessment and situation analysis must be conducted. While scenario analysis evaluates the general demands and difficulties experienced by the affected people, damage assessment teams evaluate the structural damage to buildings, infrastructure, and utilities. This data aids in prioritizing recovery and relief activities and directing resources.
- Logistics and supply chain management: These two areas must operate effectively for an emergency response. This entails securing, transporting, and distributing to the impacted areas in a timely manner, basic necessities such as food, water, medical equipment, and makeshift shelters. The quick delivery of resources to the affected people, meeting their urgent needs, and supporting response actions are ensured by effective logistical coordination (White, M. P. 2012).
- Social services and psychological support: Response and emergency management also take into account the psychological health of those who are impacted. Counseling, mental health evaluations, and trauma interventions are just a few of the psychological support services that assist people in overcoming the emotional trauma and stress brought on by the hazard occurrence. Social services help to fulfill the immediate needs of individuals affected, such as by providing food, clothing, and assistance with temporary housing.
- In order to respond effectively to emergencies, it is crucial to manage information and ensure good public communication. Making informed judgments is facilitated by the establishment of trustworthy mechanisms for data collection, analysis, and sharing by responding agencies. Providing regular updates, safety warnings, and details on the resources and services that are accessible are all examples of public communication. Accurate and transparent communication promotes confidence, informs the public, and supports a well-coordinated response effort.
- Recovery and continuity planning: Following the response and disaster phase, recovery efforts concentrate on reestablishing normalcy and reconstructing devastated communities. Planning for continuity ensures the restart of key services, aids in the restoration of infrastructure, and makes it easier for displaced

populations to return. Financial aid is given, house reconstruction is facilitated, utilities are restored, and the devastated region's economic recovery is supported.

Communities can lessen the immediate effects of risks, save lives, and set the stage for the process of recovery and rebuilding by putting effective response and emergency management systems into place. Communities can recover from the difficulties brought on by hazard events with the aid of successful coordination, prompt action, and prioritizing the needs of the impacted populations (Ceacero-Moreno, 2021).

10.6.5 Recovery and Rehabilitation

Following a hazard occurrence, recovery and rehabilitation entail the reconstruction of infrastructure and local communities. This principle underlines the necessity for long-term recovery plans that take into account social, economic, and environmental factors in addition to rebuilding physical infrastructure. Supporting damaged people, aiding the restoration of livelihoods, incorporating hazard-resistant design concepts, and fostering long-term resilience are all part of it. Recovery efforts should focus on lowering potential vulnerabilities and improving the impacted areas and their general resilience (White, 2012).

Let's examine some crucial components of healing and rehabilitation:

- Damage assessment and needs analysis: Conducting a thorough damage assessment and needs analysis is the first step in the recovery process. This entails determining the degree of physical harm to buildings, homes, companies, and public spaces. Analyzing the social and economic effects on people, households, and the neighborhood at the same time is essential. Effective resource allocation and prioritization are made possible with the use of this information.
- Infrastructure restoration: A major goal of recovery and rehabilitation is to restore essential infrastructure. Roads, bridges, schools, hospitals, water and sanitization systems, and electrical grids that have been harmed can all be repaired or rebuilt. By using hazard-resistant design and construction techniques to reduce potential vulnerability in the future, infrastructure restoration strives to increase the system resilience.
- Housing recovery: One of the top priorities in the recovery process is to address the housing requirements of affected people

and households. This entails offering alternatives for temporary housing, aiding the restoration or repair of damaged dwellings, and helping people whose homes are uninhabitable relocate. In order to increase future resilience, housing recovery programs may also involve enforcing housing laws and standards.

- Economic recovery and livelihood restoration: Supporting local businesses' recovery and livelihood restoration is essential for ensuring the community's resilience. Initiatives for economic recovery are centered on giving loans, grants, and financial support to struggling firms, assisting in the reopening of stores and markets, and encouraging entrepreneurship and job creation. To aid in the restoration of livelihoods, initiatives can also include skill development, vocational training, and assistance with agricultural endeavors.
- Support for social and psychological wellbeing: Recovery and rehabilitation initiatives focus on the social and psychological health of impacted people and communities. To assist people in overcoming the emotional trauma brought on by the hazard occurrence, this includes offering counseling services, mental health assistance, and community outreach initiatives. A comprehensive rehabilitation process requires the rebuilding of social networks, the promotion of social inclusion, and the promotion of community cohesiveness.
- Environmental restoration: Recovery and rehabilitation depend heavily on environmental restoration. It entails restoring natural ecosystems that may have been harmed or destroyed by the hazard event, including forests, wetlands, and coastal regions. In addition to improving ecological health, environmental restoration offers long-term advantages such as erosion management, flood protection, and biodiversity preservation (Ceacero-Moreno, 2021).
- Building institutional capacity and governmental organization strength is a goal of recovery and restoration activities in order to increase local institutions' resilience and readiness for dangers in the future. Programs for training, knowledge exchange, and technical help are all part of this effort to improve disaster management capabilities. The construction of resilience and sustainable recovery systems depend on effective institutional

coordination, effective governance, and community involvement (White, 2012).

- Planning for long-term resilience: Recovery and rehabilitation offer the chance to make long-term resilience plans. This entails incorporating hazard mitigation and adaptation strategies into infrastructure design, land use restrictions, and urban planning. Planning for long-term resilience emphasizes lowering risks in the future, enhancing early warning systems, taking climate change into account, and encouraging sustainable development methods.
- Collaboration and stakeholder involvement: Strong stakeholder involvement and collaboration are essential for successful recovery and rehabilitation. This entails including participation from regional organizations, companies, non-governmental organizations, academic institutions, and governmental bodies in the decision-making process. By including stakeholders, rehabilitation plans are promoted as being sustainable and context-specific, and ownership is fostered.
- Monitoring and evaluation: It is essential to track the development of recovery and rehabilitation activities in order to gauge the success of interventions and pinpoint areas in need of development. This entails monitoring important indicators, evaluating results, and getting stakeholder feedback. Monitoring and evaluation make rehabilitation efforts more in line with the changing demands of the community and serve to shape future hazard management measures.

A multifaceted, coordinated strategy for recovery and rehabilitation is needed, one that takes into account the unique circumstances and requirements of the afflicted population. These principles help communities remain resilient in the face of threats in the future by addressing physical, social, economic, and environmental elements (Ceacero-Moreno, 2021).

10.6.6 Collaboration and Stakeholder Involvement

These two concepts together make up a key hazard management principle. Governments, communities, non-governmental organizations, corporations, and different facets of society must work together to address dangers. Sharing information, resources, and expertise is made possible through effective teamwork, which also makes it easier to come up with coordinated plans of action. Promoting ownership, inclusion, and responsibility in hazard management initiatives requires including stakeholders at all levels. Let's explore the significance of stakeholder participation and collaboration in hazard management:

- Shared responsibility: Collaboration and stakeholder participation highlight the fact that hazard management is a shared responsibility amongst several organizations and authorities. It acknowledges that everyone who has an interest in risk reduction and resilience development has a role to play, including governmental entities, local communities, corporations, non-governmental organizations, and academic institutions (White, 2012). Shared responsibility encourages a team effort in which participants pool their resources, expertise, and abilities to achieve a common objective.
- Local knowledge and context: Involving stakeholders makes sure that methods for managing hazards take into account local knowledge, experiences, and viewpoints. Local communities have important knowledge about the dangers they confront, their weaknesses, and the coping strategies they use. Utilizing this knowledge by including local stakeholders improves the efficacy and relevance of hazard control techniques. For solutions to be implemented that are both culturally and contextually suitable, it is essential to comprehend the local context.
- Holistic approach: A holistic approach to hazard management is fostered by stakeholder participation and collaboration. Diverse stakeholders each provide a unique set of skills, including knowledge in the technical, scientific, social, and cultural spheres. It is possible to gain a thorough understanding of the dangers and their effects by including stakeholders from a variety of fields, including infrastructure, health, environment, disaster management, and urban planning. This all-encompassing strategy produces risk assessments that are more thorough, creative solutions, and well-informed choices.
- Improved decision-making: Including stakeholders in hazard management procedures results in more intelligent and successful decision-making. Every stakeholder contributes their own special priorities, values, and points of view, which broadens the decision-making process. Open discussion, consideration, and

reaching a consensus are encouraged by collaborative decision-making. This leads to improved outcomes and greater acceptance and support for hazard management activities by ensuring that choices are transparent, inclusive, and take into consideration the interests of all stakeholders (Ceacero-Moreno, 2021).

- Resource optimization: Through collaboration, resources and capabilities can be improved. Each stakeholder contributes unique financial, personal, technological, and logistical resources to the discussion. Working together enables resources to be more efficiently pooled and used for hazard management actions. Resource sharing, cooperative planning, and coordinated action are also made possible through collaboration, which reduces effort duplication, increases efficiency overall, and saves money.
- Stakeholder involvement fosters creativity and supports the investigation of fresh concepts and methods for hazard control. Engaging participants from various backgrounds develops creativity and promotes unconventional thinking. Innovative solutions, including early warning systems, resilient infrastructure designs, community-based adaption measures, and sustainable land-use practices, can be created by stakeholders working together to explore and develop them. Collaboration also promotes learning and the sharing of best practices, enabling participants to modify and enhance their strategies in light of lessons discovered from various contexts (White, 2012).
- Ownership and Empowerment: Including stakeholders in hazard management procedures gives them the ability to actively participate in decision- and action-making. Stakeholders become more invested in the outcomes when they have a sense of ownership and agency. As a result, there is an increase in commitment, involvement, and accountability in the application of hazard management strategies. Participation of stakeholders fosters the capacity and resilience of the local community by empowering groups and individuals to implement preventative and risk-reduction measures.
- Collaboration and stakeholder involvement among the various players involved in hazard management help to promote trust and social cohesion. Trust is fostered by open and transparent communication, active participation, and inclusive processes,

which promote collaboration. Stakeholders are more inclined to share information, cooperate, and work toward common objectives when they trust one another and the process. Due to the fact that it brings together various stakeholders and fosters respect, understanding, and a sense of shared duty, collaboration also fosters social cohesion.

- Sustainability and Effective Implementation: Involving stakeholders improves the sustainability and efficacy of hazard control programs. Stakeholders' requirements, worries, and feedback are taken into account when they are included throughout the full hazard management cycle, from planning and execution to monitoring and evaluation. The possibility of successful implementation is increased by this participatory method since stakeholders have a stake in the results and may contribute their knowledge and expertise on the ground. Collaboration also promotes a sense of ownership and long-term commitment, which makes hazard control measures sustainable after the first response stage (Ceacero-Moreno, 2021).
- Collaboration and stakeholder involvement build a culture of readiness, collaboration, and group action, which helps communities be more resilient. Engaging society's stakeholders at all levels strengthens social networks, builds social capital, and promotes a common awareness of risks and obligations. Communities become more resilient as a result of collaboration because they are better prepared to foresee, respond to, and recover from risks. Participation of stakeholders fosters empowerment and capacity building as well, empowering communities to take the initiative to lessen their vulnerabilities and increase their resilience (White, 2012).

In conclusion, cooperation and stakeholder participation are necessary for efficient hazard management. Initiatives for hazard management become more thorough, creative, and sustainable by involving stakeholders, utilizing their knowledge and resources, and fostering cooperation. Collaboration and stakeholder involvement ultimately help to create resilient communities that can deal with and lessen the effects of hazards.

10.6.7 Continuous Improvement and Learning

The final hazard management element, continuous improvement and learning, highlights the significance of learning, adapting, and improving continuously. Emerging dangers, technology improvements, and shifting societal dynamics all have an impact on the field of hazard management, which is constantly changing. The discovery of gaps and possibilities for improvement is made possible by routine evaluations, monitoring, and feedback methods. Hazard management procedures can be improved and enhanced over time by consistently learning from the past and incorporating new information and innovations (White, 2012).

The fundamentals of hazard management offer a thorough framework for comprehending, evaluating, and dealing with distinct threats. Stakeholders may successfully decrease risks, improve readiness, respond effectively to catastrophes, speed up recovery, and create resilient communities and systems by embracing these ideas. Hazard management is a constant effort that calls for cooperation, creativity, and a dedication to protecting people's safety and wellbeing in the face of potential hazards (Ceacero-Moreno, 2021).

10.7 IMPLEMENTATION OF HAZARD MANAGEMENT

The process of putting strategies and practices into place to lower risks, get ready for emergencies, and improve resilience is known as hazard management. Comprehensive planning, risk analysis, mitigation strategies, readiness, response, and emergency management, as well as recovery and rehabilitation, are all part of it. It also calls for ongoing evaluation and improvement. Thorough planning, which involves identifying risks, evaluating vulnerabilities, creating objectives, and developing coordinating channels, is the first step in hazard management execution. The likelihood and potential effects of hazards are then assessed through risk assessment, which serves as a foundation for prioritizing activities and creating mitigation solutions.

Risk-reduction strategies are used to lessen the effects of hazards. These can include structural measures like early warning systems and public awareness campaigns as well as non-structural ones like land use planning and physical interventions like infrastructure renovations. Activities that promote readiness guarantee that people are ready to deal with hazards. These include creating emergency action plans, running drills and exercises, and instructing staff. Involving the general public, encouraging resilience in both the person and the community, and forming alliances with pertinent stakeholders are also included. The reactions and emergency management phase are activated during a hazard incident. This entails mobilizing resources, organizing emergency services, and putting response strategies into action. Lives are saved, property is protected, and impacted communities are supported by quick and efficient response methods (White, 2012).

After the immediate response, recuperation and rehabilitation become the main priorities. Infrastructure restoration, help to those who need it most, and community rehabilitation are all part of this phase. Plans and policies are revised to improve future resilience after lessons learned have been evaluated. Assessing the effectiveness of hazard control activities requires ongoing evaluation. It includes tracking results, getting input, and pinpointing areas that need work. Future modifications and enhancements to plans, policies, and procedures are informed by evaluation findings. Strong leadership, efficient coordination, enough funding, and stakeholder involvement are necessary for a successful implementation. It is a continuous process that adjusts to changing community requirements, new knowledge, and increasing risks. In the face of dangers, hazard management seeks to preserve life, reduce damage, and strengthen community resilience (Ceacero-Moreno, 2021).

REFERENCES

1. Barbier, E. B., &Hochard, J. P. (2018). Land degradation and poverty. *Nature Sustainability*, *1*(11), 623-631.
2. Bellard, C., Bertelsmeier, C., Leadley, P., Thuiller, W., &Courchamp, F. (2012). Impacts of climate change on the future of biodiversity. *Ecology Letters*, *15*(4), 365-377.
3. Bortey-Sam, N., Nakayama, S. M., Ikenaka, Y., Akoto, O., Baidoo, E., Yohannes, Y. B., ... & Ishizuka, M. (2015). Human health risks from metals and metalloid via consumption of food animals near gold mines in Tarkwa, Ghana: Estimation of the daily intakes and target hazard quotients (THQs). *Ecotoxicology and environmental safety*, *111*, 160-167.
4. Chan, E. Y. Y., Huang, Z., Lo, E. S. K., Hung, K. K. C., Wong, E. L. Y., & Wong, S. Y. S. (2020). Sociodemographic predictors of health risk perception, attitude and behavior practices associated with health-emergency disaster risk management for biological hazards: the case of COVID-19 pandemic in Hong Kong, SAR China. *International Journal of environmental research and public health*, *17*(11), 3869.
5. Chowdhury, S., Kim, G. H., Bolan, N., &Longhurst, P. (2019). A critical review on risk evaluation and hazardous management in carcass burial. *Process Safety and Environmental Protection*, *123*, 272-288.
6. Collins, A. E., Manyena, B., Jayawickrama, J., & Jones, S. (2015). Introduction: Hazards, risks, and disasters in society. In *Hazards, risks,*

and disasters in society (pp. 1-15). Academic Press.

7. D'Odorico, P., Bhattachan, A., Davis, K. F., Ravi, S., &Runyan, C. W. (2013). Global desertification: Drivers and feedbacks. *Advances in water resources*, *51*, 326-344.
8. Davies, T. (2015). Landslide hazards, risks, and disasters: introduction. In *Landslide Hazards, Risks, and Disasters* (pp. 1-16). Academic Press.
9. Djalante, R., Shaw, R., &DeWit, A. (2020). Building resilience against biological hazards and pandemics: COVID-19 and its implications for the Sendai Framework. *Progress in Disaster Science*, *6*, 100080.
10. Eiser, J. R., Bostrom, A., Burton, I., Johnston, D. M., McClure, J., Paton, D., ... & White, M. P. (2012). Risk interpretation and action: A conceptual framework for responses to natural hazards. *International Journal of Disaster Risk Reduction*, *1*, 5-16.
11. Eisner, M. D., Anthonisen, N., Coultas, D., Kuenzli, N., Perez-Padilla, R., Postma, D., ... &Balmes, J. R. (2010). An official American Thoracic Society public policy statement: Novel risk factors and the global burden of chronic obstructive pulmonary disease. *American Journal of Respiratory and Critical Care Medicine*, *182*(5), 693-718.
12. Embry, M. R., Belanger, S. E., Braunbeck, T. A., Galay-Burgos, M., Halder, M., Hinton, D. E., ... & Whale, G. (2010). The fish embryo toxicity test as an animal alternative method in hazard and risk assessment and scientific research. *Aquatic Toxicology*, *97*(2), 79-87.
13. Fernández, P., &Ceacero-Moreno, M. (2021). Urban sustainability and natural hazards management; designs using simulations. *Sustainability*, *13*(2), 649.
14. Grimm, N. B., Chapin III, F. S., Bierwagen, B., Gonzalez, P., Groffman, P. M., Luo, Y., ... & Williamson, C. E. (2013). The impacts of climate change on ecosystem structure and function. *Frontiers in Ecology and the Environment*, *11*(9), 474-482.
15. Guan, W. J., Zheng, X. Y., Chung, K. F., &Zhong, N. S. (2016). Impact of air pollution on the burden of chronic respiratory diseases in China: time for urgent action. *The Lancet*, *388*(10054), 1939-1951.
16. Guerrant, R. L., DeBoer, M. D., Moore, S. R., Scharf, R. J., & Lima, A. A. (2013). The impoverished gut—a triple burden of diarrhoea, stunting and chronic disease. *Nature Reviews Gastroenterology & Hepatology*, *10*(4), 220-229.
17. Hermans, K., &McLeman, R. (2021). Climate change, drought, land

degradation and migration: exploring the linkages. *Current opinion in environmental sustainability*, *50*, 236-244.

18. Hermans-Neumann, K., Priess, J., & Herold, M. (2017). Human migration, climate variability, and land degradation: hotspots of socio-ecological pressure in Ethiopia. *Regional Environmental Change*, *17*, 1479-1492.
19. Hessel, R., Reed, M. S., Geeson, N., Ritsema, C. J., Van Lynden, G., Karavitis, C. A., ... &Witsenburg, K. (2014). From framework to action: the DESIRE approach to combat desertification. *Environmental Management*, *54*, 935-950.
20. Hoelzer, K., Moreno Switt, A. I., & Wiedmann, M. (2011). Animal contact as a source of human non-typhoidal salmonellosis. *Veterinary research*, *42*(1), 34. https://doi.org/10.1186/1297-9716-42-34
21. Hunter, L. M. (2015). Migration and environmental hazards. *Population and environment*, *26*, 273-302. doi:10.1007/s11111-005-3343-x
22. Jha, M. K. (2010). Natural and anthropogenic disasters: An overview. *Natural and anthropogenic disasters: vulnerability, preparedness and mitigation*, 1-16.
23. Koks, E. E., Jongman, B., Husby, T. G., &Botzen, W. J. (2015). Combining hazard, exposure and social vulnerability to provide lessons for flood risk management. *Environmental science & policy*, *47*, 42-52.
24. Lawrence, Timothy. (2015). Pollination and Protecting Bees and Other Pollinators. 10.13140/RG.2.1.3099.1203.
25. Levin, R., Zilli Vieira, C. L., Rosenbaum, M. H., Bischoff, K., Mordarski, D. C., & Brown, M. J. (2021). The urban lead (Pb) burden in humans, animals and the natural environment. *Environmental research*, *193*, 110377. https://doi.org/10.1016/j.envres.2020.110377
26. Liang, W., Bai, D., Wang, F., Fu, B., Yan, J., Wang, S., ... & Feng, M. (2015). Quantifying the impacts of climate change and ecological restoration on streamflow changes based on a B udyko hydrological model in C hina's L oess P lateau. *Water Resources Research*, *51*(8), 6500-6519.
27. Meier, H. M., Müller-Karulis, B., Andersson, H. C., Dieterich, C., Eilola, K., Gustafsson, B. G., ... &Schimanke, S. (2012). Impact of climate change on ecological quality indicators and biogeochemical fluxes in the Baltic Sea: a multi-model ensemble study. *Ambio*, *41*, 558-573.

28. Mirzabaev, A. ; Wu, J. ; Evans, J. ; Garcia-Oliva, F. ; Hussein, I. A. G. ; Iqbal, M. H. ; Kimutai, J. ; Knowles, T. ; Meza, F. ; Nedjroaoui, D. ; Tena, F. ; Türkeş, M. ; Vázquez, R. J. & Weltz, M. (2019). Desertification. In P. R. Shukla, J. Skeg, E. Calvo Buendia, V. Masson-Delmotte, H.-O. Pörtner, D. C. Roberts, P. Zhai, R. Slade, S. Connors, S. van Diemen, M. Ferrat, E. Haughey, S. Luz, M. Pathak, J. Petzold, J. Portugal Pereira, P. Vyas, E. Huntley, K. Kissick, M. Belkacemi & J. Malley (eds.), Climate Change and Land: an IPCC special report on climate change, desertification, land degradation, sustainable land management, food security, and greenhouse gas fluxes in terrestrial ecosystems
29. .Nicholls, C. I., & Altieri, M. A. (2013). Plant biodiversity enhances bees and other insect pollinators in agroecosystems. A review. *Agronomy for Sustainable development*, *33*, 257-274.
30. Pannure, A. (2016). Bee pollinators decline: perspectives from India. *International Research Journal of Natural and Applied Sciences*, *3*(5), 2349-4077.
31. Paton, D., Buergelt, P. T., McCaffrey, S., &Tedim, F. (2014). *Wildfire hazards, risks, and disasters*. Elsevier.gyu
32. Paul, B. K. (2011). *Environmental hazards and disasters: contexts, perspectives and management*. John Wiley & Sons. 10.1002/9781119979616.
33. Polidoro, B. A., Carpenter, K. E., Collins, L., Duke, N. C., Ellison, A. M., Ellison, J. C., ... & Yong, J. W. H. (2010). The loss of species: mangrove extinction risk and geographic areas of global concern. *PloS one*, *5*(4), e10095.
34. Prüss-Ustün, A., Vickers, C., Haefliger, P., &Bertollini, R. (2011). Knowns and unknowns on burden of disease due to chemicals: a systematic review. *Environmental health*, *10*(1), 1-15.
35. Pyšek, P., & Richardson, D. M. (2010). Invasive species, environmental change and management, and health. *Annual review of environment and resources*, *35*, 25-55.
36. Rawat, U. S., & Agarwal, N. K. (2015). Biodiversity: Concept, threats and conservation. *Environment Conservation Journal*, *16*(3), 19-28.
37. Reid, A. J., Carlson, A. K., Creed, I. F., Eliason, E. J., Gell, P. A., Johnson, P. T., ... & Cooke, S. J. (2019). Emerging threats and persistent conservation challenges for freshwater biodiversity. *Biological*

Reviews, *94*(3), 849-873.

38. Reynolds, J. F., Grainger, A., Stafford Smith, D. M., Bastin, G., Garcia-Barrios, L., Fernández, R. J., ... &Zdruli, P. (2011). Scientific concepts for an integrated analysis of desertification. *Land degradation & development*, *22*(2), 166-183.
39. Rodriguez-Lazaro, D., Cook, N., Ruggeri, F. M., Sellwood, J., Nasser, A., Nascimento, M. S. J., ... & van der Poel, W. H. (2012). Virus hazards from food, water and other contaminated environments. *FEMS microbiology reviews*, *36*(4), 786-814.
40. Saad, A. M. A., Shariff, N. M., &Gairola, S. (2011). Nature and causes of land degradation and desertification in Libya: Need for sustainable land management. *African Journal of Biotechnology*, *10*(63), 13680-13687.
41. Shi, P., Ye, T., Wang, Y., Zhou, T., Xu, W., Du, J., ... & Okada, N. (2020). Disaster risk science: A geographical perspective and a research framework. *International Journal of Disaster Risk Science*, *11*, 426-440.
42. Skogen, K., Helland, H., &Kaltenborn, B. (2018). Concern about climate change, biodiversity loss, habitat degradation and landscape change: Embedded in different packages of environmental concern?. *Journal for Nature Conservation*, *44*, 12-20.
43. Smith, K. (2013). *Environmental hazards: assessing risk and reducing disaster*. Routledge.
44. Taylor, R. G., Scanlon, B., Döll, P., Rodell, M., Van Beek, R., Wada, Y., ... &Treidel, H. (2013). Ground water and climate change. *Nature climate change*, *3*(4), 322-329.
45. Ulyshen, M., & Horn, S. (2023). Declines of bees and butterflies over 15 years in a forested landscape. *Current Biology*, *33*(7), 1346-1350.
46. Wernberg, T., Russell, B. D., Moore, P. J., Ling, S. D., Smale, D. A., Campbell, A., ... & Connell, S. D. (2011). Impacts of climate change in a global hotspot for temperate marine biodiversity and ocean warming. *Journal of experimental marine biology and ecology*, *400*(1-2), 7-16.
47. Wilcove, D. S., &Koh, L. P. (2010). Addressing the threats to biodiversity from oil-palm agriculture. *Biodiversity and conservation*, *19*, 999-1007.

48. World Health Organization. (2011). *Burden of disease from environmental noise: Quantification of healthy life years lost in Europe*. World Health Organization. Regional Office for Europe.
49. Zattara, E. E., &Aizen, M. A. (2021). Worldwide occurrence records suggest a global decline in bee species richness. *One Earth*, *4*(1), 114-123.

INDEX

H

I

J

L

T

V

W

Z

Printed in the United States
by Baker & Taylor Publisher Services